FANTASIES OF VIRTUAL REALITY

\<strong\> Ideas Series
Edited by David Weinberger

The \<strong\> Ideas Series explores the latest ideas about how technology is affecting culture, business, science, and everyday life. Written for general readers by leading technology thinkers and makers, books in this series advance provocative hypotheses about the meaning of new technologies for contemporary society.

The \<strong\> Ideas Series is published with the generous support of the MIT Libraries.

FANTASIES OF VIRTUAL REALITY

UNTANGLING FICTION, FACT, AND THREAT

MARCUS CARTER AND BEN EGLISTON

THE MIT PRESS   CAMBRIDGE, MASSACHUSETTS   LONDON, ENGLAND

The open access edition of this book was made possible by generous funding and support from the MIT Libraries.

The MIT Press would like to thank the anonymous peer reviewers who provided comments on drafts of this book. The generous work of academic experts is essential for establishing the authority and quality of our publications. We acknowledge with gratitude the contributions of these otherwise uncredited readers.

This book was set in ITC Stone and Avenir by New Best-set Typesetters Ltd. Printed and bound in the United States of America.

Library of Congress Cataloging-in-Publication Data is available.

ISBN: 978-0-262-54916-5

10 9 8 7 6 5 4 3 2 1

CONTENTS

CONTENTS

1

FANTASIZING ABOUT VIRTUAL REALITY

In 480 BCE, as part of the Second Persian Invasion of Greece, the powerful King Xerxes I of Persia commanded two feats of extraordinary technical ambition and hubris: a pontoon bridge across the Dardanelles so that his army could cross safely into Greece, and a canal across the isthmus of Mount Athos, a dangerous headland that had destroyed the Persian fleet a decade earlier.

These feats of engineering stand out in history not for their success—the bridge was destroyed by the time the army returned, and the canal collapsed soon after its only use—but for how they represent the extraordinary power and scope of the Persian state under King Xerxes I. Over six hundred ships, 1,300 anchors, and thirty kilometers of anchor rope all supported a bridge more than two kilometers long.

It does not matter that it was a disastrously bad investment. King Xerxes had the money and the will to conjure these tremendous feats of engineering, and for a brief time the citizens of Greece were forced to live in a world where these technologies existed. Such was the divine power of kings, and today, such is the divine power of big tech.

In late October 2021, Mark Zuckerberg announced on stage at his company's annual developer conference that Facebook would be rebranded as *Meta*, to reflect its ambition "to help bring the metaverse to life." Imprecisely defined, the *metaverse* is framed in his accompanying founder's letter as the successor to today's internet: part of the natural evolution of the internet "from desktop to web to mobile; from text to photos to video," to an "embodied internet, where you're in the experience, not just looking at it." This dramatic rebrand by one of the world's largest technology companies shifted massive attention toward one of the emerging technologies that Zuckerberg proposes will underpin this next paradigm of computing: virtual reality (VR).

While many commentators critiqued the rename as a self-interested (if not desperate) attempt to distance the company from its toxic reputation and numerous governance failures on its social media platforms, Facebook and Instagram, Meta's investments in VR are long-standing. Following Facebook's IPO (the then-highest valued technology IPO in US history), the company underwent a process of significantly expanding its *platform boundaries*, as platform scholars David Nieborg and Anne Helmond put it,[1] in a series of high-value corporate acquisitions, purchasing companies like Instagram (2012) and WhatsApp (2014). In 2014, Facebook acquired Oculus, a gaming-focused VR company launched through a successful Kickstarter campaign only two years prior. Since then, Meta has spent over $1 billion on related VR and augmented reality (AR) acquisitions, and it has significantly expanded its research and development arm, spending a reported $12 billion on VR and AR development in 2021 alone, with the staff deployed representing as much as one-fifth of Meta's total workforce.

To put this into perspective, the amount of money that Zuckerberg's Meta is spending to make his "metaverse" a reality each year is roughly equivalent to an entire year's research and development expenditure of businesses in Australia. This is King Zuckerberg's world—and we will have to live in it, even if only for a short time.

The purpose of this book is to help readers understand the promises and pitfalls of VR. In *Fantasies of Virtual Reality*, we critically examine the promises of VR—in entertainment, for empathy, for socializing, in education, and for enterprise—identifying the amazing opportunities that this emerging technology presents while dispelling tech boosterism and naive optimism. In going beyond the fantasies of VR and delving into the sometimes messy and problematic realities, we highlight how VR—and thus the metaverse—is one of the most data-hungry digital sensors we're likely to invite into our lives in the next decade, with enormous potential for exclusion, manipulation, and harm.

Our use of the term *fantasies* in this book is a play on the concept of the social imaginary—via George E. Marcus's *technoscientific imaginaries*[2]—that acknowledges how our collective perceptions of the power of technology and the role that technology may play in the future shape the ways that technologies are developed, deployed, and regulated. Our approach to fantasies is also attuned to questions of how the imagination (and development, and use) of VR adheres to particular social logics. In this way, we also owe a debt to the thinking within the *social construction of technology theory*—a theoretical approach developed in the 1980s out of science and technology studies—which would suggest that technological development and use is driven by the way that particular social actors interpret technology.[3] *Fantasies* also intends to capture the fantastic, and often unfounded, nature of the modern VR imaginary. It is here that we find some foothold in cultural theorist Lauren Berlant's notion of fantasy—idealized and often (knowingly) unrealistic or impossible expectations of life, collectively internalized.[4] Through going beyond describing the experience of VR, and taking seriously the material, political, and ethical aspects of VR, we will provide a critical account of VR's capabilities that goes beyond the common tech-booster discourse that is widespread today, identifying what is possible, what is misleading, and what is just plain fantasy.

Our book is organized around the most pervasive and central fantasies that developers, investors, and boosters have for VR: in gaming, for empathy, for enclosure, for violence, and for data collection.

One of the reasons that VR's boosters perpetuate these fantasies is that there is something genuinely *fantastic* about VR. In our teaching, we've had the opportunity to put hundreds of students through their very first experience using VR, and nearly all students emerge genuinely excited about how immersive and *real* it felt. *Virtual presence*—the sensation of feeling physically embodied in a virtual space, as if you're actually there—is a phenomenal feeling. It underpins many of big tech's promises about VR. It primes us to accept claims like those made by VR filmmaker and entrepreneur Chris Milk, who describes VR as the "ultimate empathy machine," or Jeremy Bailenson, founding director of the Stanford Virtual Human Interaction Lab, who claimed in 2015 that "we are entering an era that is unprecedented in human history, where you can transform the self and experience anything the animator can fathom." Against the backdrop of this potential, what use are concerns about technological limitations, or the sexism and exclusion built into its design?

Despite this promise, VR has remained on the precipice of widespread adoption for nearly a decade. Why? To critique the notion of immersion, as we do, is not to say that VR cannot engender immersion or that it is not more immersive than 2D media, but to argue that uncritically accepting these advantages as somehow core or fundamental to the medium is fraught with danger, and assuming this property can limit the potential. As we demonstrate, going beyond the fantastic and critically understanding the underlying affordances of VR will be necessary in understanding how and where it might make the most difference.

Another theme throughout this book is frustration with the uncritical treatment of VR in tech journalism and academia. In comparison to other widely analyzed and critiqued emerging technologies like artificial intelligence (AI) or crypto, VR was rarely discussed before Zuckerberg's rebrand. In part, we argue, this is because it was typically thought of as a gaming and entertainment technology, not a serious one with the potential for far-reaching impacts upon society. This is, unfortunately, false. Each generation of VR has expanded the amount of data that it collects about the user and their environment, and the potential deployments of VR in workplaces, in schools, and by the military have the potential for far-reaching harms. Even more frustratingly, the prominence of Zuckerberg's futuristic metaverse vision attracts the overwhelming majority of critique, not the harms it presents today. Irrespective of whether the metaverse takes hold, the scale of his investment in making it a reality means that some form of it is an inevitability, even if only for a short time. Understanding its potential now, before it comes entrenched, is critical.

2

FANTASIES OF GAMING

INTO THE GAME

VR emerged from science fiction as a potentially real gaming fantasy in the early 1990s, when the Japanese video game company Sega publicized a VR add-on for their popular Sega Genesis console at the 1993 Consumer Electronics Show. Promising entry "into the game" for $200, Sega VR (figure 2.1) was framed by marketing slogans that capture a similar fantasy that VR presents players today, with engrossed players in the promotional video exclaiming that it's "hard to remember it's just a game," that "it's like being there," or that "it's like you've got a movie living in your head." Despite only giving a small number of journalists controlled access to a working prototype, consumer enthusiasm was high, and Sega VR was dubbed the product of the year by *Popular Science*.

However, by 1994, promotion had stopped for the Sega VR, and it was removed from the company's release schedule. Sega claimed that the innovative gaming device was canceled because the VR effect "was so realistic, it could potentially cause injury to children who played it," indicating that players would forget they were in a virtual world. The real reason was motion sickness: in 2015, Sega's CEO at the time confirmed that "almost everybody got sick . . . it caused severe motion sickness. Other people got severe headaches."[1]

VR so often leads to motion sickness because its central objective is to trick the sensorial capacities of the body to make it feel like you're actually in the virtual environment. This is what creates the incredible sensation of presence in a VR world. However, when there is a conflict between our perception (what we see) and our proprioception (what we physically

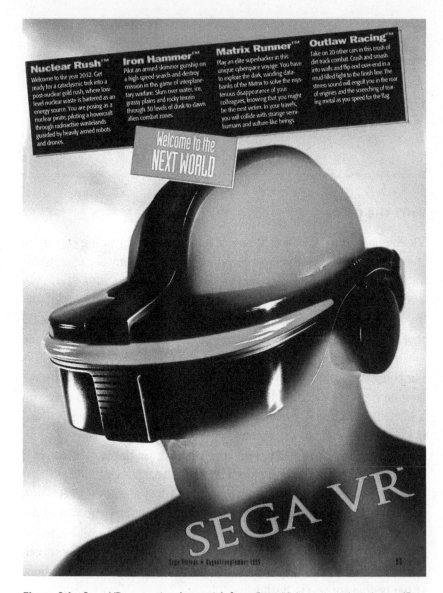

Figure 2.1 Sega VR promotional material, from *Sega Visions* magazine, August/September 1993.

feel), our bodies—trained by a million years of evolution—assume that we have been poisoned and that vomiting up that poison is the best solution. VR developers have come up with fascinating techniques—such as making the display black-and-white when moving the head quickly—to reduce the impact of motion sickness, but this vulnerability still found in modern VR highlights how closely interwoven the VR interface is with the human body.

The challenge that faced the Sega VR was that it ran as a peripheral on the Sega Genesis, a home console with a 7.6 MHz CPU and 64 KB of RAM, capable of displaying up to sixty-one colors at once. While competitive for a video game console at the time, the Sega VR could display only four to twelve frames per second—versus the 90–120 frames per second on modern headsets—which is "much too slow to keep up with users turning their head."[2] This lag meant that the eyes saw motion that didn't match up with what the vestibular system in the inner ears felt, causing motion sickness.

Only four games were ever in development for the Sega VR system, and with the cancellation of the hardware, none were ever released.[3] In 2020, Rich Whitehouse—head of digital conservation at the Video Game History Foundation—secured access to the source code for one of these games, *Nuclear Rush*, and was able to create an emulator to play the game on a modern HTC VIVE VR headset.[4] Set in the year 2032, "where low-level nuclear waste is bartered as an energy source," the player is "posing as a nuclear pirate, piloting a hovercraft through radioactive wastelands guarded by heavily armed robots and drones" (see figure 2.2). The player's head movements are separate to the movement of their hovercraft, allowing them to navigate the wasteland to find fuel while targeting enemy planes and tanks by aiming the crosshair in the middle of the screen by turning their head. Whitehouse described the game as "a lot more playable than I'd expected."

Concurrent with the Sega VR, which was ultimately unsuccessful at disrupting the home console market, in the early 1990s there was a flourishing industry of VR devices in video game arcades, theme parks, and malls. The most significant of these included the Virtuality system in London in 1991[5] and the VR-1 VR amusement park attraction released by Sega at the Joypolis theme park in Japan in 1994.[6] In contrast to the

Figure 2.2 A screenshot from the unreleased Sega VR game *Nuclear Rush* (courtesy Video Game History Foundation).

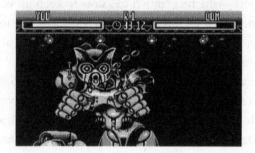

Figure 2.3 A screenshot from the Nintendo Virtual Boy game *Teleroboxer*.

$200 price tag Sega aimed to place on the Sega VR, these systems involved significantly more advanced hardware and could cost between $55,000 and $75,000. Closer to theme park experiences than to domestic console gaming, arcade-based VR remained a popular attraction in Japan through the early 2000s (Virtuality became insolvent in 1997), further contributing to the fantasy of modern of VR as an innovative and advanced gaming experience.

One proto-VR headset that followed the Sega VR was Nintendo's Virtual Boy, released in 1995. Like the Sega VR, the Virtual Boy used two screens and a parallax effect—that is, each eye received a slightly different image—to create the illusion of 3D. However, the Virtual Boy could only display red and black images and did not feature head tracking (figure 2.3). Instead, the Virtual Boy was attached to a stand on a table or desk, with the screens more like a pair of affixed binoculars at a tourist-viewing station than like a modern VR headset. So although motion sickness wasn't an issue—there's no proprioceptive conflict if you're not moving your head—reviews describe significant eye strain from the low-resolution screens and neck pain from players holding their heads at awkward viewing angles.

While technically not VR,[7] the Virtual Boy is significant in the history of VR because of the impact that its failure had on the broader VR industry and how it established VR gaming as a desirable technological aspiration. Through the early 1990s, Nintendo's dominance in the US domestic game market was challenged by Sega, with the sixteen-bit Sega Genesis outselling the Super Nintendo Entertainment System in the United States. One of the key ways Sega accomplished this was through appealing to an older male teen audience with "edgy," hypersexual, and misogynistic advertisements, a marketing strategy that other companies adopted and that subsequently had the effect of contributing toward the toxic male-oriented "gamer" identity that characterizes much of contemporary games culture.[8] In his account of the failure of the Virtual Boy, Stephen Boyer suggests that "the Virtual Boy in many ways is a pure appeal to the older male teen audience that in the mid-1990s was thought to desire edgy content and boundary-pushing technology."[9] So though the Virtual Boy was discontinued in June 1995 as Nintendo's worst-selling console, shipping only 770,000 units, it defined and attempted to sell a fantasy about VR gaming that was fully enmeshed with hypermasculine notions of the *gamer*.

Like Sega, the experience Nintendo described in its advertising and promotion for Virtual Boy—and the experience consumers expected based on VR's depiction in film and fiction—simply wasn't possible with the technology they had available. In a 1994 press release, Nintendo president Hiroshi Yamauchi is quoted as stating that the Virtual Boy "will transport game players into a 'virtual utopia' with sights and sounds unlike anything they've ever experienced." In the United States, TV advertisements described "a 3D game for a 3D world," and elsewhere Nintendo described the Virtual Boy as "immersing players into their own private universe." These adverts—more so than the Virtual Boy itself—had the effect of constructing VR as both a technological and experiential ambition, similar to the pursuit of ever-increasing graphical fidelity that has characterized video game development over the past twenty years. It is the idea that more pixels and more polygons will inevitably lead to a better gaming experience, a gaming fantasy predicated on a narrow idea of what good gaming experiences look like.

Stephen Boyer points out that the red-only display—chosen because red LEDs were cheaper—"eliminated any possibility of the virtual world passing as a new form of reality, except as one that was generally unpleasant," noting further that "the games that were ultimately released for the system did little to produce a new world: most were just 2-D games with 3-D effects."[10] The inconsistency between the marketing of the Virtual Boy and its technological capabilities contributed heavily to the failure of the console, a failure that killed off the idea of domestic VR in the 1990s and kicked off what is commonly referred to as *VR's long winter*, when few products or innovations were made. After all, if Nintendo—one of the most innovative companies in games—couldn't launch a successful VR product, who could?

VR'S REEMERGENCE

If we take posts on the Meant to Be Seen (MTBS) VR forum as a starting point, then VR's reemergence began in August 2009, with a project by then-seventeen-year-old Palmer Luckey to make a head-mounted 3D VR gaming device. Luckey had attempted to make his first iteration of a VR headset—a yellow fiberglass helmet—in his parents' garage, and over the subsequent eighteen months he continued iterating, posting in June 2010: "I have gotten halfway done on 5 or 6 now, only to realize that I could greatly improve, then scrap and start over."[11] As the story goes (according to Luckey), he spent much of his time as a teenager tinkering with technology and repairing iPhones for cash—cash that he'd spend on a collection of VR headsets, many of which were from the 1990s. Amassing this collection (supposedly the largest private collection in the world)[12] allowed him to learn about the technology, and his search for a missing part to a 1992 VR headset saw him hired as a lab technician at a University of Southern California research lab in 2011,[13] working on military-funded research that explored topics such as the potential for VR to help treat post-traumatic stress disorder.

Luckey continued to post about his progress on the MTBS forums, which imbued Oculus with a particular "hacker aesthetic,"[14] mirroring in many ways the ethos of technological ambition that surrounded the Virtual Boy. Oculus's first headset—the Rift—is regularly described in reports and histories as having been "cobbled together" from disassembled smartphone components, perpetuating this early mythology for the device. Luckey would post updates on each prototype, and when he announced his intention to launch a crowdfunding campaign on Kickstarter, he regularly participated in discussions to emphasize that it was not for "regular gamers": "I would rather not have regular gamers buying this not understanding what it is and the limitations, and then giving it all kinds of bad reviews!" (April 2012); and "I want it very clear on the Kickstarter that this is for the DIY/hacker/enthusiast crowd, not a mainstream product" (June 2012). Forum members were also given the opportunity to preorder directly from Luckey before the Kickstarter was opened, and the device was specifically described as a *developer kit*, not a consumer headset. Based on interviews with early VR adopters, media studies scholar Maxwell Foxman suggests[15] that this helped Luckey build a community of practice of hobbyist modders developing for the Rift, which quickly helped establish it as the default VR platform.

One early reader of Luckey's posts was John Carmack, lead programmer on several of history's most influential first-person shooter games, including *Doom* and *Wolfenstein*. Concurrent with working on the remaster of *Doom 3*—which included support for 3D displays—Carmack was experimenting with the different VR headsets that were available at the time.[16] After connecting through his forum posts, Luckey sent one of his prototype VR headsets to Carmack, who took it to E3—a large gaming expo—in 2012, catalyzing an avalanche of interest in the project. Although described at the time by the Eurogamer website as "a publicity stunt to promote a re-release,"[17] Carmack's involvement put Oculus and VR on a trajectory toward a particular kind of gaming fantasy: the pinnacle of the hyperviolent, high-graphical-fidelity fantasy that characterizes hardcore games in the first-person shooter genre that Carmack—far beyond anyone else—has pioneered. As Carmack states in a 2012 interview: "All that we've been doing in first-person shooters, since I started, is try to make virtual reality. Really, that's what we're doing with the tools that we've got available. The sole difference between a game where you're directing people around and an FPS is, we're projecting you into the world to make that intensity and that sense of being there and having the world around you. . . . So the lure of virtual reality is always there."[18]

Luckey too was interested in this potential for gaming. As his prototypes improved, he described his ambition toward recreating the Virtuality arcade experience that had been attempted in the 1990s, and in the Kickstarter update announcing his $16 million Series A funding, he emphasized that he "got into VR because it seemed like the obvious path to the best possible gaming experience."[19] But, as Leighton Evans summarizes, "the intentionality and politics of designers, programmers and manufacturers should be thought of as written large into the design, materiality and experience of VR."[20] As we'll see, VR was pitched to—and co-opted by—a particular genre of gaming in which its promises were most immediately obvious: hardcore gaming.

OCULUS RIFT: STEP INTO THE GAME

Echoing the precise language that had been used to promote the Sega VR nearly twenty years prior, Luckey's newly formed company, Oculus, launched its Kickstarter in August 2012. Proudly "designed for gamers, by gamers," the headset was pitched exclusively as a gaming device, one "taking 3D gaming to the next level": "Oculus Rift is a new virtual reality (VR) headset designed specifically for video games that will change the way you think about gaming forever. With an incredibly wide field of view, high resolution display, and ultra-low latency head tracking, the Rift provides a truly immersive experience that allows you to step inside your favorite game and explore new worlds like never before."[21] In a campaign video, Luckey grounds his desire to create the Rift in the idea that "there was nothing that gave me the experience that I wanted, the *Matrix*, where I can plug in and actually be in the game."[22] A glowing endorsement from Carmack preceded a list of similar endorsements from other (male-only) heavyweights in the hardcore gaming community, including Cliff Bleszinski, lead designer of the *Gears of War* series; Gabe Newell, president of Valve (which develops games like *Half Life*, *Portal*, and *Team*

Fortress and operates Steam, the primary digital distribution platform for computer games); and David Helgason, founder of game development software company Unity Technologies. As with earlier attempts at VR, Oculus's fantasy for VR gaming remained aspirational to hardcore gamer cultures' core values of wholly immersive, violent, technologically advanced, and challenging gameplay.[23]

The emphasis on a "truly immersive experience" mirrors the discourses around Sega VR and the Virtual Boy. But what, precisely, is meant by *immersion*? VR researcher Mel Slater defined immersion in 1996 as something quantifiable, the "extent to which the computer displays are extensive, surrounding, inclusive, vivid and matching."[24] VR gaming discourses often adopt this conceptualization of immersion, drawing on concepts like embodiment and presence—which are so powerfully and immediately felt when using VR for the first time—to bolster their position, but an approach like this fails to acknowledge what makes games so immersive. *Tetris*, for instance, excels at providing an immersive experience, but it is driven by entirely different experiential phenomena than immersion in a rich role-playing game such as *Skyrim*. Game studies scholars Laura Ermi and Frans Mäyrä propose a multidimensional definition for immersion in games: sensory immersion (the audiovisual aspects), challenge-based immersion (similar to flow), and imaginative immersion (becoming absorbed in the story or world, similar to becoming "absorbed into a good novel").[25] While they noted that VR can provide "the purest form of sensory immersion,"[26] it is these other aspects that contribute to the overall immersiveness of games—and as we'll discuss later, without all these dimensions, VR experiences can fall flat.

Crucially, the Oculus Kickstarter was not for a consumer device but for a *developer kit*, a prototype version of the device for developers to use in order to integrate it with their existing games (although each version included a copy of Carmack's *Doom 3: BFG Edition*, which could be played on the Rift without any programming knowledge). This is common in software development as it enables developers to create for the consumer device before it is released. Despite this emphasis, the Oculus Kickstarter captured the imagination of an enormous number of players and developers, who quickly committed to support what the games press were calling "the future of gaming."

Eventually, 9,522 backers pledged a little over $2.4 million to support the Kickstarter campaign. In 2013, once Dev Kit 1 (DK1) started shipping to backers, direct sales continued at a rapid pace, far beyond the DIY/hacker crowd Luckey had originally intended. Oculus shipped sixty thousand DK1s by March 2014, and by February 2015, the company had shipped over one hundred thousand successor DK2 headsets. This was largely possible due to two significant injections of venture capital (VC): $16 million Series A funding in June 2013, and $75 million Series B funding in December 2013 (led by Andreessen Horowitz). While a consumer edition was still years from becoming available, VR gaming finally looked poised for mainstream success.

It is at this point in the history of VR's emergence that Meta (then Facebook), and Mark Zuckerberg, becomes the main character, acquiring Oculus for $3 billion in March 2014.[27] This made Oculus the first billion-dollar company to emerge from a Kickstarter campaign. As we discuss at length in chapter 3, this acquisition was one of several high-profile acquisitions following Facebook's 2012 IPO (the then-highest valued tech IPO in US history),[28] although at the time the purchase was often characterized in reports as an extravagance for the post-IPO cash-rich Zuckerberg driven by his interests in gaming and technology, rather than an obvious integration into the existing social platform.

In any case, the acquisition enraged many of those in the gaming community and those who had backed the original Kickstarter. Although this occurred before some of its most egregious scandals, Facebook was already an unpopular platform with the tech-enthusiast community, associated more closely with data collection, surveillance, and social media oversharing than with gaming. As Foxman summarizes, "This direction unsettled initial users of the Rift who feared the loss of the platform's gaming aspects as well as their input in the product."[29] Other concerns highlighted the ethics of having taken money from a Kickstarter campaign only to be acquired for $3 billion less than two years later, and how such a large company as Facebook might stifle innovation and competition in the VR market. Capturing the mood of the gaming community, *Minecraft* creator Markus Persson (who later that year sold his company to Microsoft for $2.5 billion) quickly announced that talks about bringing *Minecraft* to Oculus were cancelled, adding, "Facebook creeps me out."[30]

The vitriolic reception to the acquisition is grounded in VR's gaming fantasy, by that stage firmly established as wholly immersive and hardcore gameplay. If Facebook was associated with gaming, it was with casual social media games like *Farmville* and *Bejeweled*: not the types of games that industry veteran John Carmack was associated with and that backers had supported Oculus to make. Facebook was antithetical to what the VR gaming fantasy had developed into, and the acquisition was widely framed as Luckey "selling out" to Zuckerberg and "betraying" the community that had supported him. If Oculus had instead been acquired by Microsoft or Sony, large media and tech companies already clearly invested in catering to the hardcore gamer community, the reaction would not have been as rancorous. Although Facebook (as Meta) has gone on to invest billions in VR gaming—a level of investment highly unlikely to have happened if Oculus had remained independent or been acquired by another company—the unbridled enthusiasm expressed for VR during its rapid reemergence quickly tempered as it became clear that there remained many obstacles for delivering the Matrix-like experience that Luckey had sought to create.

THE GAMER BODY

VR is distinct from other media because of the extremely close coupling between the body and the VR system.[31] VR causes motion sickness when there is a conflict between our perception (what we see) and our proprioception (what we physically feel). Our bodies, trained by millions of years of evolution, can react quite violently to this kind of conflict. Estimates vary, but as many as 20 to 40 percent of people who try VR report motion sickness when using it.[32] Of course, some VR applications are worse than others: roller-coaster experiences are essentially designed to replicate the experience of a real roller coaster, motion sickness included, while more gentle ones rarely induce discomfort. Developing for VR, then, is a practice of developing for the body, and one of the consequences of VR's gaming fantasy is that it has developed for a particular kind of body: the gamer body.

The critical point here is that, in understanding VR and who is excluded from it, we have to focus on the real, not the virtual. As scholars like Daniel

Golding suggest, "The dominant 'image' of virtual reality is not as its advocates might suggest, impossible fantasy worlds fully realized through virtual reality technology, but the apparatus and the body in an unwieldy alliance."[33] Modern VR, and our fantasies of it, are primarily instantiated through media like YouTube reaction videos and news outlet reports. Here, the focus is not on the disappearance of the real but—as Golding points out—on three key elements that shape the image of modern virtual reality: the VR system, the body of the user, "and the gendered nature of that body."[34] Luckey himself, the hacker community he first sought patronage from, and the types of "hardcore gamer" games that came to characterize VR's gaming fantasy, normalized VR as "yet another domain for white, middle-class men."[35] As Foxman puts it, "In order to flourish, Oculus tapped into the passion and even ideologies of existing 'gamer' culture without addressing associated issues surrounding gender and misogyny."[36]

GENDERED REALITIES

One of the most prominent examples of exclusion in VR is against women. This is in both the design of VR hardware and in the virtual worlds we're asked to inhabit when we use VR. Various studies and reports have suggested that women experience motion sickness from VR more than men.[37] One explanation for this is that there are two key techniques that the brain uses to deduce depth: motion parallax and shape from shading. *Motion parallax* is based on the apparent size of the object; if it gets bigger, we deduce that it is getting closer to us. *Shape from shading* is more complex and is based on how the shading of an object changes very slightly when we move. Social media scholar danah boyd found that Oculus devices rely heavily on motion parallax, which just so happens to be the proprioceptive technique favored by men.[38] "In the real world," boyd notes, "both these cues work together to give you a sense of depth. But in virtual reality systems, they're not treated equally,"[39] leading to a much less comfortable experience. VR's overwhelmingly male development (and research)[40] community ultimately designed a device that suited the needs of their bodies, but not the needs of women's.

Depth perception is not the only element of the VR system in which this kind of gendered exclusion happens. Technology journalist Adi

Robertson wrote in 2016 about her experiences trying VR prototypes that were "designed for a range of body types that I—a fairly normal-sized woman—cling to the very lower edge of": haptic jackets that are far too large to work, headsets that barely tighten enough to fit on her head, and an eye-tracking system that failed because she was wearing mascara.[41] Robertson describes the experience as "a literal, concrete inability to use technology in the way it's intended to be used, simply because you're outside an artificially-skewed norm."[42]

A concrete example of this is *interpupillary distance* (IPD), the distance between the centers of the user's two eyes. Lining up the center of the eye with the lens within a VR headset is important for comfortable and nonblurry use. The adjustable IPD range in the Quest 1, for instance, is 58–72 mm, which is "best" for 99 percent of men but only 93 percent of women.[43] The Rift S, which doesn't allow adjustment, is even more problematic, only suitable (according to Meta) for users with an IPD between 61.5 mm and 65.5 mm. While this covers roughly the same proportion of men and women (46 percent of men, versus 43 percent of women), the distribution of this fit is much closer to the mean male (64.67 mm) than mean woman (62.31 mm; see figure 2.4), meaning that more women are much further outside the range of bodies for which the headset was designed.[44] Recent research by Kay Stanney and colleagues found that IPD was found to be the primary driver of gendered differences in motion sickness.[45] The worse the IPD nonfit is, the worse the motion sickness

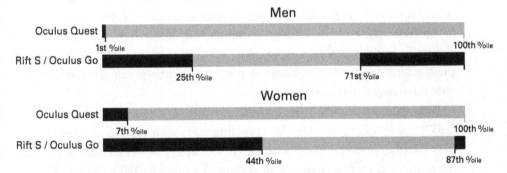

Figure 2.4 Using data from the ANSUR II Anthropometric Survey, UploadVR highlighted how the IPD range of the Rift S headset specifically caters toward male users, which results in an increased likelihood of motion sickness for women. *Source:* Heaney, "Data Suggests."

is.[46] These data sets are public, and their use is standard in the design of computer hardware. Women are being excluded simply by not being envisioned as users of the system.

GENDERED VIRTUAL REALITIES

Even when these barriers baked into the hardware of VR can be overcome, women remain excluded from equal participation with VR because of online harassment. One of the most prominent early pieces of writing about VR harassment is by UX researcher and author Jordan Belamire, who outlined her experience of sexual assault in the form of a "virtual groping" in social VR archery game *QuiVr*, describing the experience as something that was very real in terms of its felt effects.[47] Belamire's experience, as surveys of VR users' social experiences point out, is not uncommon. For surveyed users of VR in 2018, 49 percent of female and 36 percent of male respondents reported experiencing some form of sexual harassment.[48] In describing her sexual assault in 2016, Belamire notes, "The virtual groping feels just as real. Of course, you're not physically being touched . . . but it's still scary as hell."[49] The account Belamire provides is reminiscent of earlier observations about virtual harassment—specifically, Julian Dibbell's well-known 1993 article "A Rape in Cyberspace," in which a user, expert in the affordances of a text-based virtual environment, nonconsensually mediated sexual encounters between other players and themselves—something no less harmful despite there being "no bodies touched."[50] VR-based sexual harassment is particularly problematic due to how VR affords the user an immersive, haptically, auditorily, and visually rich experience. As we've shown in this chapter, these affordances are often taken as positives, but they can also make virtual harassment acutely traumatic.

Following Belamire's 2016 assault and article about the incident, *QuiVR*'s developers responded by providing users with more power over their (virtual) personal space, writing: "If VR has the power to have lasting positive impact because of that realism, the opposite has to be taken seriously as well."[51] Changes made to *QuiVr* following the publication of Belamire's account included a "personal bubble" feature that means that other players "fade out" when they reach out to grab or touch another, but not all games implement this feature.

Raising serious concerns that the safety of these digital spaces have not been improved since these issues became widely known in 2016, a 2023 study by the Center for Countering Digital Hate found that users encountered sexually explicit harassment, racist abuse, and misogyny within five minutes in 20 percent of *Horizon Worlds* environments.[52] Nonprofit advocacy group SumOfUs (now Ekō) recently reported that a researcher was in *Horizon Worlds* for less than an hour before she experienced sexual assault on the platform, with one user simulating sex with her avatar, while another user watched and drank a bottle of vodka. They then gave her the bottle, saying: "Here you're going to need more of this."[53] VR blogger RoybnzReality has described how she is frequently "swarmed" by other players when she enters public lobbies in VR.[54] As VR has become more accessible, children have also become victims of this kind of harassment. Divine Maloney has documented extensive derogatory comments toward minors, particularly young girls,[55] noting elsewhere that, "children are particularly noticeable on social VR platforms" due to voice chat, and they are commonly subject to targeted harassment.[56]

In response to media reportage on incidents like these, Nick Clegg, president of global affairs at Meta (and former UK deputy prime minister) argued, "We wouldn't hold a bar manager responsible for real-time speech moderation in their bar, as if they should stand over your table, listen intently to your conversation, and silence you if they hear things they don't like." However, as Kate Clark and Trang Le suggest, it is disingenuous to equate social VR spaces to managing a real-life bar, as public spaces accessed via VR sit between digital spaces and real-life spaces and therefore need to be moderated differently than both physical spaces and other, flat-screen digital spaces.[57] Legal scholar John Danaher argues that some instances of what he terms *virtual sexual assault* are real "because virtual sexual assault can have real world consequences"—that is to say, real harm—"and there are some grounds for thinking that certain aspects of sexual activity are social, as opposed to physical, in nature."[58]

A critical issue here is that there are clashing social norms about how people should behave in online spaces, and what constitutes harassment.

For instance, in gamer culture, *teabagging* refers to the act of crouching repeatedly over another player's body—in most games, after the player has been killed. It was popularized in the competitive play of the first-person shooter game *Halo*, which had an unusual feature allowing killed players to view their body—and the teabagging act of dominance[59]—for a few moments after their death: just long enough for them to be taunted by the victor. Numerous games have since designed in the ability. VR users encultured in gaming communities like these see these forms of avatar-to-avatar harassment as appropriate ways to interact online, in VR included. For instance, in 2016, games YouTuber PewDiePie (who for several years ran the most subscribed to nonbrand channel on YouTube) released an eight-part video series reviewing the HTC VIVE. The first of these is simply titled "Teabagging in VR." By not addressing these issues head-on—and explicitly soliciting hardcore gamer community members as early adopters of VR—Meta, Valve, and other early VR developers have fostered a discriminatory online VR culture.[60]

VIRTUAL REALITY AS A DISABLING TECHNOLOGY

This pattern is similarly repeated in VR's focus on embodiment, specifically in embodying the *normate* body—the socially constructed, ideal image of the body[61]—in VR. *Embodiment* is "the ensemble of sensations that arise in conjunction with being inside, having, and controlling" a virtual body.[62] For people with disabilities—whose bodies do not meet VR's expectation of what a body should look like—the focus on embodiment excludes them from equal participation in VR. To adopt a perspective from the wider field of critical disability and media disability studies, it is crucial here to understand disability not in terms of deficit (what the body can't do and thus why it can't engage with technology in a particular way) but rather from the perspective that technology can *disable* the user, by failing to be designed in a way that accounts for users' diverse requirements.[63] A web interface, for instance, that cannot be read by a screen reader for visually impaired users has the effect of disabling those users.

While some in disability studies are hopeful for the promise of VR's multisensory aesthetics for more sensorially inclusive media experiences,[64] more recent work in human-computer interaction (HCI) has productively identified how contemporary VR systems reinforce ableism.[65] In 2017, a report on survey research from a partnership of Lucasfilm's ILMxLAB and the Disability Visibility Project gave an account of the experience of people with disabilities in using VR, pointing out various ways that VR interfaces are "disabling" by virtue of their design. Elsewhere, Able-Gamers, a disability advocacy group focused on increased accessibility for video games, has presented a comprehensive breakdown of accessibility issues in VR. These include heavy emphasis on motion controls, requirement of very specific body positioning, or a privileging of the visual and gestural (with less attention to accessible audio).[66] While disabled users often feature prominently in the slick advertisements put out by VR companies attempting to shape our impression of what could be possible with the technology, many of those users would not be able to comfortably play many of the most popular games available today.

One clear example of this is the lack of control over avatar height with the Meta Quest. Through the Quest's Insight system—and its ability to track position and orientation—the device is able to situate the user as accurately as possible in the virtual environment. This means that if you crouch in real life, your avatar crouches in the game. But for wheelchair users and people with limited mobility, such an approach makes many Quest games unplayable. Interfaces are often designed to be only within reach of the standing user, and the sitting user's view is rendered at the crotch height of the virtual nonplayer characters. Seated mode, where it does exist, is designed for the comfort of the normate body (an opportunity for rest), rather than the inclusion of those for whom sitting is a necessity. Even games advertised as being playable in seated mode assume a level of mobility that renders them unplayable for some users, asking players—who might, for instance, be confined to a hospital bed—to bend down to pick a virtual item up off the floor, which they may not be able to do.

In short, the kinds of bodies and mobilities that are rendered machine-readable are based on (ableist) design-level assumptions about what bodies are and what they can do. As one frustrated VR user on the Oculus forums put it, VR does not "recognize me as an adult human being"[67]. If these types of options were configured at the level of the technology, VR could be inclusive of a wider variety of users, but they are currently up to each individual developer. Virtual environments are entirely virtual; they can be rendered in ways that are inclusive. Instead, the assumptions that Meta and VR designers make about their users are disabling, excluding people from the promises of the platform.

With the concept of the metaverse being used to situate VR as a more mundane, everyday technology—something that can be integrated into everyday life—the case of disability is a key example of how Oculus still very much emboldens the same fantasy of VR as a libertarian, identity-free, and disembodied fantasy.[68] If VR is to become central to interaction and participation in contemporary societies, then the stakes of exclusion are significant.

A TRAGIC FANTASY

In 2017, Mark Zuckerberg stated his goal of reaching one billion users in VR within the next ten years, a target that remains impossibly far-fetched six years later. But why hasn't VR gaming taken off?[69] The Quest lines of headsets retailed for (a heavily subsidized) $299.99, far cheaper than an Xbox or PlayStation console. Although Meta doesn't release sales figures, estimates of the number of devices sold range from four to fifteen million,[70] and unlike the Sega VR or Nintendo's Virtual Boy, modern headsets actually live up to the advertising hype. In part, we can return to the body to understand some of VR's inherent limits. Headsets are heavy, and uncomfortable when worn for prolonged periods. Meta recommends that people take a break every thirty minutes, and even as regular users whose bodies are well suited to VR, we feel the need to follow this advice. There is a huge gulf between the unfettered enthusiasm most people have after using modern VR for the first time and its broader commercial adoption.

What we want to conclude the chapter with here is the argument that VR's gaming fantasy—of advanced technology, wholly immersive and aspirational to the core values of hardcore gamer culture—is ultimately flawed, and this flawed fantasy has held back the potential of VR for gaming. As we've charted, VR gaming has long been situated as something technologically advanced and enmeshed with the hypermasculine notions of the hardcore gamer. As such, VR development and marketing has focused on graphical realism and replicating the ideals of hardcore games on a new platform. These two fronts have been VR gaming's downfall.

BUT CAN IT RUN CRYSIS?

Over the past three decades, *AAA game development*—a term used in the games industry to signify high-budget games distributed by large publishers—has driven the massive expansion of computing power in consumer gaming devices. Particularly in PC gaming, part of what made a game "hardcore" was the computing power needed to run it at the maximum settings—that is, with the most detailed and textured graphics available. As Stephanie Boluk and Patrick Lemieux write, "The computer and video game industry has been caught up in a graphical arms race: a dogged pursuit of ocularcentric spectacle culminating in the hypertrophy of the visual economy of games."[71] Video game marketing—constrained by the noninteractivity of screenshots, trailers, and billboards—focused on these advancements in graphical fidelity, shaping the discourse around what a "good game" should look like. The 2007 first-person shooter game *Crysis* had maximum setting system requirements that pushed the limits of what was possible on even enthusiast PCs at the time. This was immortalized in the "But can it run *Crysis*?" meme, a tongue-in-cheek way to query a PC's power and, in turn, emphasize the importance of graphic fidelity to gaming culture.

Gaming culture's attention to graphics is based in the enormous and highly visible changes to game graphics over the past thirty years, which have contributed to significant improvements in player experience. While comparable advancements were made in other areas, such as in game audio and in the interactivity of virtual environments, it was graphical realism that became most closely entwined with the inflated concept of immersion. As we've already discussed, *immersion* is loosely and poorly defined and, as the example of *Tetris* shows, is not based solely on graphical realism. For PC and console gaming, players and developers are increasingly realizing that there are diminishing returns in improving the look of near-photorealistic avatars; in some cases, they're honing back the realism to avoid the uncanny valley effect of not-quite-perfect human avatars.

As VR emerged on the horizon in the 2010s, it was quickly adopted into gaming discourses about computing power. Hardware manufacturers began slapping VR Ready stickers on high-end PCs, even before consumer VR devices were readily available. PC-based VR often called for system requirements that required a PC costing $2,000 to $3,000 because rendering a virtual environment correctly—that is, to avoid motion sickness—required a powerful PC. This introduced yet another barrier to the widespread adoption of VR—but it reflected a continuation of the values that underpin the "But can it run *Crysis*?" meme: more computing power and graphical realism means a better gaming experience. VR game developers—caught up in, and constrained by, gaming culture's valorization of graphical realism and focus on sensory immersion—have typically sought to recreate the graphical realism of flat games in VR. But first VR's *Crysis*-like demand on consumer hardware and later the constraints of mobile VR[72] have meant that VR games have struggled to demonstrate an edge in graphical realism. The one exception to this rule is likely *Half Life: Alyx*, the only genuinely AAA-level, VR-only game. It was developed by Valve, which invested an estimated $50 million into the game's development.

One game that sidesteps this competition—and has subsequently become one of the most celebrated VR games—is *Super Hot VR* (figure 2.5). Unlike *Alyx*, the graphics in *Super Hot* are strikingly basic. The player's physical environment is minimalist and untextured, and opponents are closer to early 2000s PS2 graphics than to the photorealistic avatars of many modern games. In the PC version of *Super Hot*, time moves when the player moves their avatar. In *Super Hot VR*, time moves when the player moves *their body*. The attention in *Super Hot's* gameplay is not on ocularcentric sensory immersion but on what genuinely makes VR distinct: embodiment. The effect of this mechanic is a superhuman-like control of time manipulation, providing the player the "Matrix-like" experience that Palmer Luckey described in the original Oculus Kickstarter campaign, deftly contorting the body to evade slow-moving bullets while dispatching enemies with an empowering ease. This immersion-through-proprioception *feels good*, and it provides a VR-version of gaming power fantasy, but it had to shed gaming's fantasy of photorealism to do so.[73] By attending to the unique affordances of the medium, *Super Hot VR*

Figure 2.5 A promotional still from the video game *Super Hot VR*, which remains one of the top-selling VR games five years after its release due to its unique experience of time manipulation through body movements. *Source: Super Hot VR* (https://www .terminals.io/games/superhot-vr).

provides an experience worth donning a headset for in the recommended thirty-minute intervals, and it consequently remains one of VR gaming's top-selling titles five years after its release.[74]

FARMVILLE VR

Another movement in the games industry that has shaped VR's gaming fantasy is the dichotomy of "casual" and "hardcore" gaming that underpins the modern gamer identity. As scholars like Carly Kocurek have demonstrated, masculinity has long been central to how games have been represented and promoted.[75] In the 1990s, Sega pioneered the older male teen focus of games through "edgy," hypersexual, and misogynistic advertisements, shaping today's toxic gamer culture.[76] However, as the audience of players expanded through the late 1990s and early 2000s—such as through massively multiplayer games that attracted an older audience, party games like *Dance Dance Revolution*, and consoles such as the Nintendo Wii that reasserted gaming as a social and family-oriented activity—the dichotomy of casual and hardcore games entered the gaming lexicon as part of an (ongoing) debate about what games (and players) matter more.[77] As Shira Chess summarizes, "Traditionally, 'hardcore' describes games that are difficult to learn, expensive, and unforgiving of mistakes and that must be played over longer periods of time. Conversely, casual games can be learned quickly, are forgiving of mistakes and cheap or free, and can be played for either longer or shorter periods of time, depending on one's schedule."[78]

Scholars have extensively critiqued this hardcore/casual dichotomy as fundamentally flawed and a deeply inadequate way to understand games and gaming. Casual puzzle games, for instance, are often played in hardcore ways.[79] But distinctions like these, and debates over which games are "real games" and which are not,[80] are instances of boundary work that seek to invalidate certain types of games (e.g., narrative-based games, or games that do not perpetuate power fantasies) and certain types of players (e.g., women and LGBTQIA+ people). The harms of this culture can be most clearly found in the #GamerGate movement, an internet-based harassment campaign against diverse players and game developers which began in 2014.[81] Despite this—and to reiterate the point by Foxman—"in order to flourish, Oculus tapped into the passion and even ideologies of existing 'gamer' culture without addressing associated issues surrounding gender and misogyny."[82] The values of this community consequently oriented VR gaming toward using this new medium to create hardcore experiences.

For instance, one of Oculus's original and most visible proponents was John Carmack, lead programmer on some of the most pioneering game franchises of all time, including *Doom* and *Wolfenstein*. These first-person shooters closely map to hardcore gaming ideals: they have a limited focus on story; they're (hyper)violent, fast-paced, and difficult; and they feature some of the goriest scenes in games. In the same way that VR's gaming fantasy has been detrimentally entwined with the pursuit of photorealism, VR's gaming fantasy has been co-opted by these hardcore values that ultimately limit the medium. They lack mainstream appeal and valorize experiences that simply aren't as appealing in VR as they are on a flat screen.

In a discussion centered on the design of *Half Life: Alyx*, designers Greg Coomer and Robin Walker explained that VR changes the way that people interact with virtual environments.[83] As Coomer explained, "People are slower to traverse space, and they want to slow down and be more interactive with more things in each environment. It has affected on a fundamental level how we've constructed environments and put things together." This is not "because of some constraint around how they move through the world," Walker notes; instead, "it's just because they pay so much more attention to things and poke at things." Environments in VR games are much denser; on PC, they feel small, but in VR, they feel big. *Super Hot VR*'s designers similarly described how reusing levels designed for PC felt wrong, requiring the spaces to be completely redesigned.[84]

This in part explains why few games originally designed for flat screens and then ported to VR have been successful. This difference changes how we experience content: the rapidly paced hyperviolence best characterized by *Doom* is simply sensory overload; the "intensity of being there"—one of Carmack's aspirations—is unappealing in VR. Most hardcore games are simply hypermasculine power fantasies, providing a sense of mastery over the virtual environment through violent domination. In VR, unforgiving games are unpleasurable. Most of us aren't that coordinated, and we can't play for extended periods of time in VR as it is physically exhausting. This isn't to say that violence and feeling physically under threat while playing isn't appealing in some instances (e.g., VR horror is a popular genre), but that trying to cater to the values and ideals of hardcore gamers has meant that the true opportunities for the medium have not yet been fully unlocked.

In contrast, the VR game *Beat Saber* is a prime example of a game that might be derided as casual, if it weren't the best-selling VR game of all time.[85] *Beat Saber* is a music-based rhythm-matching game, a hybrid of *Dance Dance Revolution*, *Guitar Hero*, and *Fruit Ninja*. In time with (typically) electronic music, a score of red or blue boxes streams toward the player. Armed with two neon swords—commonly described as lightsabers—the player must strike these boxes in the correct direction, denoted by a subtle white arrow. Striking a box releases a note in the accompanying song, resulting in an experience that is half playing an instrument and half dance. Well-patterned songs draw on sweeping movements and rhythms that create a player experience reminiscent of the "gestural excess" that is characteristic of play on the Nintendo Wii.[86] Like *SuperHot VR*, *Beat Saber's* appeal is immersion through embodiment, also achieved by disregarding VR's gaming fantasy of hardcore experiences. With each song being, well, song length, *Beat Saber* supports a shorter, casual mode of engagement that isn't pleasurable because it is difficult or competitive, but simply because playing a song *feels good*. In 2019, *Beat Sabre* became the first VR game to sell more than a million copies, and it was purchased by Meta for an undisclosed amount.

Gaming in VR has been subjected to a vicious, self-reinforcing cycle wherein VR developers create hardcore games, which appeal to a certain kind of hardcore gamer user, whose purchasing habits in turn drive further development of those kinds of games, and not others. Attempts to penetrate this feedback loop have been met not just with the hostility of VR's online gaming culture, appropriated from gamer culture at large as VR developers sought to appeal to an established market, but also the hostility of VR's hardware that only caters to narrow and normate conceptualizations of the body. As a result, the scope of VR games remains narrow, and oblivious to the kinds of games that might come to define the genre.[87] Perhaps ironically, then, the one thing that could save VR gaming is the one thing VR enthusiasts decried the most when Facebook purchased Oculus in 2014: *Farmville VR*.

3

FANTASIES OF EMPATHY

While the modern VR resurgence was grounded in the hardcore gamer fantasy that Oculus proposed, other potential uses for VR quickly became popularized. One of the most significant of these was the technosolutionist *virtual reality for good* movement, an idea that the greater capacities for immersion and presence in VR could be used like a Band-Aid for some of the world's most significant problems, like human connectedness, climate change, sexual harassment, or police violence. Our point in this chapter isn't that VR can't be used for good, but that the idea that VR could solve society's problems in a new way was an attractive proposition for newly minted VR companies and for the Silicon Valley technosolutionist ethos. This chapter explores these fantasies of VR being for good or for change, and asks, What good? What change? And it asks critically if VR can really bring about change at all.

To begin answering these questions, we'll first need to take a trip to the zoo.

VIRTUAL REALITY IN THE ZOO

Unfortunately, an experience that most of us have shared in our lives is standing at the edge of a leafy animal enclosure at the zoo and being unable to see anything. Eventually, zoo visitors will inevitably give up, or someone will spot something: a shadow of an animal concealed behind a rock or a blur in the treetops above. This experience is—for the benefit of animal welfare—by design.[1] The architecture of the modern zoo privileges naturalism and immersion, designing enclosures to replicate the natural habitats of animals: placing African plants and scenery in lion enclosures, for example, or creating rainforest-like environments for orangutans.[2] But thanks to the propensity of wild animals to do what they can to hide from people, and the millions of years of evolution that gave them that advantage in their natural habitats, this can make zoo animals quite hard to see.

While great for animal welfare, this experience is not great for modern zoos' secondary aim as conservation organizations: encountering animals in zoos has been shown to motivate significant proconservation behavioral change.[3] This was a problem that I (Marcus) thought VR could potentially be a solution to. Since 2014, along with colleagues in the Interaction Design Lab at the University of Melbourne, I have had the opportunity to work with Zoos Victoria on the design of experimental digital technologies to enhance animal welfare and the conservation experience.[4] We partnered with Melbourne-based VR company PHORIA—which later produced the award-winning VR nature documentary series *Ecosphere* in partnership with Oculus and the World Wildlife Fund (WWF)—to explore how VR could be used in the zoo to enhance zoo visits and their conservation education outcomes. This project, and our findings, were subsequently published in the *Journal of Zoo and Aquarium Research*.[5]

Under the close supervision of zookeepers, we spent a day experimenting with filming inside animal enclosures at Melbourne Zoo with a stereoscopic 180-degree camera. When viewed on a VR headset, 180- and 360-degree footage—and the impact of that footage—can be genuinely incredible. A stereoscopic camera films with two lenses, which simulates our natural binocular vision, from which we get depth perception. When viewed in a VR headset, this footage is projected onto a dome, allowing the viewer to look around that point, and the inclusion of depth perception in that view gives a strong sensation of physical presence in the scene. It is as though you yourself are standing where the camera was affixed: in this case, standing inside a red panda enclosure about 15 cm from a small platform covered in fresh fruit.

When we recorded this footage, Mishka—who normally spends most of the day high in the treetops, almost invisible to guests—slowly climbed down, tentatively approaching the food while looking directly at the camera. The zookeeper assisting us speculated that the two binocular lenses on the camera might appear like eyes to Mishka, explaining why she looked at it so directly. When viewed in VR, it feels like Mishka is looking at *you*; wary of *you*; and *you* become conscious of your presence as an intruder, even though you *know* it's not real. Mishka eventually grows comfortable with the headset, and after a few minutes sitting on the platform eating the fruit she leans over the edge and begins sniffing the camera. Sniffing *you*. I showed one of my friends, Grace, the footage, and she raised her hand as if to touch the red panda without realizing it, and even leaned back in her chair as Mishka leaned over, much to the amusement of everyone looking on. Rather than being a soft, cute, elusive animal, high in the trees, VR gives an entirely new way of encountering Mishka; a way in which you can't help but notice how sharp her claws are. Suddenly, the extremely cute becomes wild and dangerous.

This is the transformative potential of VR that Oculus sought to fund through its VR for Good creators' program. Generations of nature documentaries have leveraged incredible filmmaking and the power of narrative to motivate care for animals in the wild, but I've never physically felt under threat watching *Planet Earth* (only existentially). With no easy way to get the camera into the treetops with the red panda, we decided to make a VR experience combining footage shot inside the penguin enclosure with behind-the-scenes footage of a zookeeper as they prepared fish with vitamins and mineral supplements to keep the penguins healthy. The narrative of the VR video connected with a significant Bubbles not Balloons conservation campaign about the impact of plastic pollution on wild seabirds.

There were two things that surprised us when we evaluated this experience with guests. The first was how profoundly social the experience was when a person was introduced to the scene. Despite being a linear video, visitors described a sensation of feeling like they had personally met the zookeeper, as though it was a one-on-one experience. It was often possible to tell what part of the video someone was in because people would start nodding during the scene in which the zookeeper starts speaking to the camera about penguin husbandry. More than just physical presence, then, the media richness of VR video can create a sense of social presence. In the context of conservation education, and other proposals for using

VR for good, this suggests that VR video might simulate the experience of being spoken to directly, a mode of delivery that is highly effective for behavioral change but often impractical to deploy at scale.

The second surprise that emerged in interviews with visitors after they had watched the video was the idea that the linear, noninteractive video was participatory. At its most basic, this was because VR requires the viewer to look around the scene, affording a choice of where to direct attention. But more strongly than this sensation of choice to look around was the sense that the experience was collaborative, as though the viewers themselves had participated in and helped in the acts of preparing the fish and then subsequently feeding the penguins. Viewers felt more involved in what was going on, describing the sensation as having "felt like I was sitting there feeding them." What this example shows is how depth perception and sensory immersion, which lead to these very real sensations of physical presence and social presence, can facilitate crucially different ways of engaging with and experiencing media content. Feeling physically and socially present in a narrative changes the way we experience that narrative. For the case of the zoo, there's clearly positive potential in VR substituting for up-close animal encounters and augmenting the at-a-distance encounters in ways that don't negatively impact welfare.

The purpose of overviewing our research in the zoo has been twofold. First, we wanted to introduce some of the qualities of a VR experience that make it so attractive to, and receptive to, the *VR for good* community, for those readers who haven't experienced VR firsthand. Second, we wanted to make clear that we're not against VR as a technology; we're just as excited about it as the boosters we critique. Our research—as well as the research of many others—has established that VR is qualitatively different than 2D media experiences and that it has a fascinating and wonderful potential to entertain, educate, and persuade.

In putting hundreds of students through their first ever encounter with VR in our teachings at the University of Sydney, it's clear that there is something *there* with VR. A potential that feels untapped. But this way that VR can be extraordinarily, surprisingly, and profoundly different from flat media lends credibility to the claims that tech boosters make about a potential for VR that is unverified, fraught with issues, and easy to be co-opted to serve other interests.

THE ULTIMATE EMPATHY MACHINE

[VR is] a machine, but inside of it, it feels like real life. It feels like truth. And you feel present with the world you are inside, and you feel present with the people that you are inside of it with. When you are sitting there in . . . [Sidra's] room watching her, you are not watching it through a television screen, you are not watching it through a window, you are sitting there with her. When you look down, you are sitting on the same ground as she is on. Because of that you feel her humanity in a deeper way. You empathize with her in a deeper way.

—Filmmaker Chris Milk, 2015[6]

The positive potential of VR for connecting us with others was popularized via Chris Milk's 2015 TED Talk on VR as the "ultimate empathy machine." In it, Milk argues that VR allows the viewer to step through the window that film provides and become a part of the virtual world, with transformative potential.

Milk's talk was widely influential, and the idea that VR affords a fundamental capacity for building empathy has been taken up widely in education, film, and various Silicon Valley start-ups. Oculus was quick to jump into this discourse, deploying $1 million in 2016 to fund the VR for Good initiative, with specific reference to Gabo Arora and Chris Milk's VR film *Clouds over Sidra*. The 2016 initiative funded ten VR films, partnering creators with nonprofits, and it was followed up in 2017 by a significantly expanded $50 million fund to develop "non-gaming, experiential VR content" to spread awareness about meaningful causes "while also highlighting the transformative potential of VR."[7] HTC also launched a VR for Impact initiative in 2017, and other VR companies and Silicon Valley philanthrocapitalists quickly followed, funding art residencies at Sundance, VR/AR tracks at South by Southwest (SXSW), and prizes at the Cannes film festival, all centered on a fantasy that VR unlocked a new solution for solving the world's most intractable problems.

To acknowledge that VR has the potential to be used for good is crucially different than claiming that it is some form of ultimate empathy machine, a fantasy about VR that quickly emerged following Facebook's acquisition of Oculus. The term *empathy machine*—initially coined by Roger Ebert to capture the great promise of cinema[8]—caught on as a way to describe VR and its transformative potential following the influential 2015 TED talk by VR entrepreneur and filmmaker Chris Milk. As Sam

Heft-Luthy points out, "Emerging media forms, throughout their history, have long been theorized as extending the human ability to imagine or connect with the inner life of another being."[9] However, the fantasy of VR's potential for generating empathy consumed much of the popular discourse around the technology.

Milk is a VR entrepreneur who cofounded VR/AR company Vrse in 2014 (rebranded as Within in 2016), initially producing primarily documentary features in partnership with organizations like Vice News, the *New York Times* (which famously provided over a million Google Cardboard VR headsets to its subscribers in 2015), and the United Nations (UN).[10] With UN Creative Director and Senior Advisor Gabo Arora, Milk (listed as the project's creative director) released *Clouds over Sidra* in 2015. The eight-minute, 360-degree video depicts life in the Za'atari refugee camp in Jordan, and is "narrated by" Sidra, a twelve-year-old Syrian refugee. The film leverages VR's capacity for physical immersion to give the camp a sense of scale, alternating between the cramped and crowded spaces of her home and school, and the seemingly endless landscape of repeating refugee tents in what has grown to be the largest Syrian refugee camp in the world, now home to nearly eighty thousand refugees. In one shot, dozens of young children run up to and surround the camera and hold hands, while "Sidra" narrates that there are more children than adults in the camp. The film ends with a call to action, leading to a link that no longer works.

Clouds over Sidra premiered in Davos, Switzerland, at the World Economic Forum (WEF), an international NGO and lobbying organization comprised of billionaire investors and business leaders. The film was widely used there and elsewhere by the UN in 2015 in its fundraising efforts for the Syrian refugee crisis, with claims praising its effectiveness, suggesting that "preliminary evidence has shown that VR is twice as effective in raising funds."[11] Milk's 2015 TED talk, titled "How Virtual Reality Can Create the Ultimate Empathy Machine," describes and evangelizes *Clouds over Sidra* and the potential for VR as follows:

And that's where I think we just start to scratch the surface of the true power of virtual reality. It's not a video game peripheral. It connects humans to other humans in a profound way that I've never seen before in any other form of media. And it can change people's perception of each other. And that's how

I think virtual reality has the potential to actually change the world. [VR] is a machine, but through this machine we become more compassionate, we become more empathetic, we become more connected, and ultimately we become more human.[12]

The conceptualization of VR as an ultimate empathy machine is built upon the idea that VR "permits one to see through another's eyes, embodying their experience, thus 'empathising' with them."[13] VR evangelist Jeremy Bailenson describes this capacity through the saying of walking a mile in someone else's shoes,[14] and this connects to the fantasy of VR experiences being psychologically real that we will discuss in chapter 5.

In a Facebook post announcing the acquisition of Oculus, Mark Zuckerberg gushed:

The incredible thing about the technology is that you feel like you're actually present in another place with other people. People who try it say it's different from anything they've ever experienced in their lives. . . . This is really a new communication platform. By feeling truly present, you can share unbounded spaces and experiences with the people in your life. Imagine sharing not just moments with your friends online, but entire experiences and adventures. . . . Virtual reality was once the dream of science fiction. But the internet was also once a dream, and so were computers and smartphones. The future is coming and we have a chance to build it together.[15]

This fantasy about VR remains at the forefront of Zuckerberg's belief in VR and the metaverse. In his founder's letter announcing the change of Facebook's name to Meta, Zuckerberg states plainly that "the defining quality of the metaverse will be a feeling of presence."[16]

WHY EMPATHY?

Before we consider the flaws in this fantasy about VR, it is worth considering for a moment why this fantasy was so quickly taken up and regurgitated across the VR ecosystem. Marina Hassapopoulou makes the point that VR (particularly in 2015–2016, when the empathy machines discourse was at its peak) is an extremely expensive technology, with VR headsets, VR-capable computers, and VR film equipment all costing thousands of dollars. As a result, VR, the topics it focuses on, and the conversations that happen around it cater "to the privileged few who can afford a smartphone and a VR headset and/or are able to attend the site-specific events and festivals"—such as the WEF meeting—"where these films are screened."[17] This, Hassapopoulou argues, "explains why there is a large segment of VR production dedicated to building 'empathy' toward—rather than a two-way conversation with—the underprivileged by narrating their stories in VR."[18] This is to say that—irrespective of the effectiveness of VR for empathy—empathy emerged as a dominant use of VR and as the dominant way in which VR was evangelized because the idea that VR could *solve* an intractable problem like empathy was specifically attractive to the types of people involved in VR discourse, funding, and production at the time: Silicon Valley's culture of technological solutionism.

Writing in *Studies in Documentary Film*, digital cultures scholar Mandy Rose argues that a parallel reason that VR has been taken up so rapidly by nonfiction documentary makers is because of the "symbiosis" between technology and content development in relation to VR, particularly at this early period.[19] Oculus was quick to jump into the ultimate empathy machine discourse, deploying $1 million to fund its VR for Good initiative in 2016 with specific reference to Gabo Arora and Chris Milk's VR film *Clouds over Sidra*. This level of investment in small-scale, independent filmmaking was significant. But the effect was that these fantasies of VR for empathy became recycled into a general concept of *VR for good* as a way for the big tech companies investing in VR to validate their investments and to construct VR as a technology with a general, mainstream audience, beyond the limited audience of gamers.

VIRTUAL REALITY IS NOT AN EMPATHY MACHINE

This fantasy of empathy is ultimately flawed. It relies on claims about vision, about empathy, and about the framelessness of VR that are false. We can unpack these fundamental flaws in the empathy fantasy theoretically, but also through practical examples.

Earlier in his ultimate empathy machine TED talk, Chris Milk uses the language of *frames*, and makes a claim to VR's framelessness against existing media:

But then I started thinking about frames, and what do they represent? And a frame is just a window. I mean, all the media that we watch—television, cinema—they're these windows into these other worlds. And I thought, well, great. I got you in a frame. But I don't want you in the frame, I don't want you in the window, I want you through the window, I want you on the other side, in the world, inhabiting the world. So that leads me back to virtual reality. . . . you feel your way inside of it. It's a machine, but inside of it, it feels like real life, it feels like truth. And you feel present in the world that you're inside and you feel present with the people that you're inside of it with.[20]

This is the faulty logic that underpins the ultimate empathy machine discourse. To present VR as a frameless medium is to claim, as Francesca Vaselli summarizes, that VR is "able to bypass the process of mediation, thus reproducing reality in a non-mediated manner—that is, reproducing the world exactly as it is."[21] For Milk, this means that your encounter with Sidra in the Za'atari refugee camp is the *experiential equivalent* to having actually traveled to Jordan and actually sat with Sidra on the floor while she told you her story. For Zuckerberg, this speaks to a broader potential—captured in his post announcing the acquisition of Oculus— that VR might allow people to share "not just moments . . . online . . . but entire experiences and adventures."[22] Indeed, such logic also underpins the rhetoric surrounding the metaverse, where virtual worlds might come to fully replace our day-to-day interactions.

The claim to VR's framelessness, or immediacy, is patently false. Just because you can see a scene in 360 degrees does not mean that the scene is not subject to mediation; that it is real; that it is "truth." To frame a scene is not only just to select where to point the camera, but also what and who is captured by that camera. Like traditional film, VR film is mediated in its choice of subject matter, in its editing, in its production, and in its display. Not to mention the fact that these scenes are linear and noninteractive.

Bimbisar Irom's critical analysis of humanitarian VR films about refugees helps unpack how the medium of VR is still subject to the same "constraints of ideology and power hierarchies"[23] that are evident in other representational tools, such as film. Irom analyses two VR films, *Clouds over Sidra* and *For My Son*, and the non-VR film *Another Kind of Girl*. *For My Son* focuses on "the professional difficulties faced by the refugees when they transition from camp,"[24] while *Another Kind of Girl*—like *Clouds over Sidra*—tells the story of a young female refugee in the Za'atari refugee camp. All three documentaries employ familiar tropes and stereotypical images, from biblical concepts to the vulnerable child, and suffer the challenge of how to address the invisibility of refugees when their voices are only heard after they pass through ideological frames. As Irom concludes, "The experiences afforded by cutting-edge immersive technology very much remain grounded as cultural products."[25] A greater sense of presence does not remove these constraints, and at worst may even conceal them from view.

Yet one of the reasons that this fantasy about VR became so widely popularized is because VR does afford a unique sensation of presence. As we described in the example of our zoo research, it does feel as though the red panda is really looking at *you*. Watching a scene in VR does evoke a response that feels like how we might respond in real life, and a response that is different from how we respond to other media like 2D video. Kate Nash highlights this in suggesting that VR carries "an inherent moral risk: the risk of improper distance," which has the effect of suppressing or diminishing any actual critical response to the documentary content. The term *improper distance* is introduced via the work of scholar Lilie Chouliaraki, who defines it as communication practices that "subordinate the voice of distant others to our own voice and so marginalize their cause in favour of our narcissistic self-communications."[26] Chouliaraki points to examples like celebrity humanitarianism and post-television disaster live blogging that privilege "the voices of the West over the voices of suffering others."[27]

VR has the potential to be similar to these forms of communication practices when it invites "self-focus and self-projection rather than a more distanced position that allows for recognition of distance between the self and other."[28] Or, as Nash succinctly puts it, "Occupying the point of view of another is no guarantee of moral engagement because there are many ways to occupy this position."[29] For the supremely privileged, watching *Clouds over Sidra* at the WEF, for instance, is to create a sense of false proximity with the refugee rather than (necessarily) encourage the viewer to think about their own role in the crisis and what can be done about it.[30] These "spurious feelings of empathy as knowledge" are, as Lisa Nakamura points out, "false," and "toxic" to combating the very problems VR for good ostensibly seeks to solve.[31] To relate back to our zoo example, there are numerous environmental-focused VR experiences that attempt to place the viewer in the perspective of a cow going to slaughter, a turtle being impacted by global warming, or a gorilla whose habitat is being destroyed. Particularly when coupled with the exciting newness of a VR experience, the hyper-reality and presence of VR presents an aesthetic mode of engagement, "inviting a contemplation of the scene as a tableau vivant or spectacle rather than a painful reality."[32]

There is perhaps no better example of this improper distance than Mark Zuckerberg's "tone-deaf live-stream" in the immediate aftermath of the devastating Hurricane Maria in late 2017, immediately ridiculed as "part disaster tourism, part product promotion."[33] With Rachel Franklin, the head of Facebook's social VR team, Zuckerberg sought to demonstrate the social features of Facebook Spaces (a short-lived social VR tool), which allowed multiple users to view 360-degree footage together. Zuckerberg talks excitedly with Franklin about how "a lot of people are using spaces . . . to go places that it wouldn't be possible to necessarily go or definitely would be a lot harder to go in real life." Appearing as his cartoonish avatar, Zuckerberg and Franklin "teleport" to a 360-degree video that NPR had released to convey the damage that Hurricane Maria had caused to Puerto Rico and surrounding islands. With the video playing, conveying the apocalyptic damage of Hurricane Maria, Zuckerberg shows off the features of the software to "look around" the video. As he stumbles to describe the damage, momentarily pausing in a way that suggests he doesn't know the name of the hurricane that had just struck Puerto Rico, he further evangelizes his VR product: "This is one of the things that is really magical about virtual reality, is that you can get the feel that you're really in a place . . . it feels like we're in the same place, and we're making eye contact and we're talking to each other." At this point in the video, as their avatars hover over streets and houses still flooded a week after the hurricane—with water, according to the NPR video, that is contaminated by sewage—Franklin and Zuckerberg awkwardly and enthusiastically high-five (see figure 3.1).

After extolling the work that Facebook was doing to help keep people safe after natural disasters (and the money that Facebook had raised and donated), they look around the scene a little longer before Franklin almost quietly confesses, "It's crazy to feel like you're in the middle of it." After this briefest of moments contemplating how the footage makes

Figure 3.1 Screenshot from Mark Zuckerberg's live stream of Facebook Spaces.

them feel about potentially *being in* Puerto Rico, Zuckerberg asks, "Do you want to go teleport somewhere else?" to which Franklin replies with a giggle, "Yeah, maybe back to California?"

Widely mocked and criticized as tone-deaf for using VR to push products at the site of a dreadful hurricane,[34] Zuckerberg's live stream perfectly reflects VR's inherent moral risk. At the time of the live stream, 88 percent of the island was without power, and close to a million people still lacked access to clean water. As he stated in the live stream—and in many of his public statements about VR—his perception was that the technology was able to make him "feel that you're really in a place," but this feeling for Zuckerberg does not, and cannot, resemble the experience of someone now homeless, without power or water, because of this hurricane. He remained, throughout the experience, a billionaire sitting in a comfortable Menlo Park office, capable of easily teleporting away (later in the live stream, they teleport to the surface of the moon). Despite VR's illusion of framelessness, who we are when we don the headset fundamentally frames our experience of VR, in a way that shapes and constrains the empathetic potential of VR.

Worse, though, we see how VR in this example centers the viewers' privileged experience of "being in" the space (as Franklin puts it, "It's crazy to feel like you're in the middle of it") rather than the experience of the people of Puerto Rico *literally still experiencing it.* Marina Hassapopoulou consequently situates VR documentaries as a new form of dark tourism, the transformation of disaster sites into mass commodities. Framing these encounters as a form of tourism connects them to the notion of the *tourist gaze*, a specific mode of engagement that tourists have with the places they visit, framed by the journey to, and from, the purposes of the journey, and the anticipation and mediation of that place.[35] Understood in this way, then, "the prospect of empathy [from VR] becomes transformed into other emotions such as relief, curiosity, morbid fascination, and, perhaps at best, a solipsistic form of gratitude [*so glad I'm not/wasn't there*]."[36] Everywhere we look; VR is framed.

After the media backlash, Zuckerberg responded in a reply to a comment on the video: "One of the most powerful features of VR is empathy. My goal here was to show how VR can raise awareness and help us see what's happening in different parts of the world. I also wanted to share the news of our partnership with the Red Cross to help with the recovery. Reading some of the comments, I realize this wasn't clear, and I'm sorry to anyone this offended."[37] As Daniel Harley points out, "An implication of Zuckerberg's response is that VR is still the solution: if we were there in VR with him, we might understand the 'sense of empathy' he claimed to feel. But Zuckerberg's presumption of empathy does little to justify the classed, raced, and gendered voyeurism of this kind of Western, risk-free, disaster tourism."[38]

One of the final challenges for this fantasy is the fact that empathy, as a concept, is a highly nebulous one, and the distinctions between other similar emotions such as pity, sympathy, and compassion are poorly recognized in this kind of popular discourse.[39] Mark Andrejevic and Zala Volcic note that Milk's preoccupation with framelessness reveals an assumption that the barrier to empathy is mediation itself—that is, that "we have to make an imaginative leap, assisted by tools of representation to get a sense of another's experience."[40] Milk's gesture to framelessness is to claim that if technology can remove the remediation of other media and reproduce the experience itself, viewers will not need to make this imaginative leap. However, in its historical usage, *empathy* literally refers to this imaginative leap. As John Ellis argues, to recognize distant others as persons involves both proximity and distance, "the imaginative attempt to feel what they are feeling and the simultaneous knowledge that they are them and we are us."[41] We cannot "know" the experience of another. Clichés like walking a mile in someone else's shoes neglect the fact that we do not experience empathy for our own situation; it is an entirely *social* emotion. As Grant Bollmer points out, empathy requires us to go beyond the limits of our own knowledge, and many VR empathy experiences are actively working against this as they focus on collapsing the self-other distinction via the presence-feeling capacities of VR.[42] No matter how effective VR is at replicating someone else's sensory experience, there is always the unbridgeable gap between the subjectivity of the user and the other.

A FANTASY OF WORKPLACE EMPATHY

One of the ways in which the empathy fantasy has, more recently, been recycled and regurgitated in corporate and VC-funded spaces is in the widespread use of VR for *workplace empathy*: sexual harassment training; diversity, equity, and inclusion (DEI) initiatives; intercultural competence development; and so forth. Many of these uses directly recycle the concepts Milk put forward in his ultimate empathy machines argument. Start-up Vantage Point, which creates VR experiences for sexual harassment training, is "founded with the belief" that "while technology can cause apathy, immersive technology can drive empathy and fundamentally make the world more human." In its 360-degree videos, participants witness actors being sexually harassed and are harassed themselves. SkillsVR claims that its Diversity, Inclusion, and Belonging training "allows participants an opportunity to respond to an emotionally charged virtual situation and to self-reflect upon their sense of curiosity and empathy towards the virtual characters they encounter." VECTRE sells its Perspectives VR platform on the basis that "it allows users to virtually experience what someone from an underrepresented identity might experience in the workplace." Mursion, a VR start-up that uses virtual avatars that are controlled live by human actors,[43] offers soft skills training centered on "sensitive topics such as unconscious bias and microinequities at work," and it organizes a yearly Actionable Empathy Symposium. VR company

Strivr—which we also discuss in chapter 6—has numerous soft skills training products, including a collaboration with the Lucile Packard Children's Hospital at Stanford that seeks to teach physicians "empathy the immersive way," allowing them to practice end-of-life conversations with parents of a terminally ill child in VR. Pervasively, these companies employ the metaphor of walking a mile in someone else's shoes in their promotional materials, with Strivr, for instance, claiming that its VR customer service training "allows associates to walk in the shoes of their customers so they're better prepared to provide excellent, empathetic service."

The potential of VR for empathy in these use cases, and the claims these start-ups make about their effectiveness, are routinely amplified by the technology and mainstream presses. Headlines like "Can VR Teach Us How to Deal with Sexual Harassment?" (*Guardian*), "Diversity Training Steps into the Future with Virtual Reality" (*Washington Post*), and "How Virtual Reality Is Tackling Racism in the Workplace" (CNN Business) parrot the claims that these founders make about the transformative effectiveness of VR over traditional training methods, despite little evidence that it works at all. Despite this, they've seen significant interest from venture capital, with VR training companies having raised over $100 million in funds since 2016. Oculus/Meta even joined this wave of enthusiasm, launching VR for Inclusion: Women in Tech in 2020, "an immersive training experience to help allies better understand what it's like for women to navigate the workplace," with accompanying reflection guides for individuals and groups. With its pervasive culture of sexual harassment and sexist design (as discussed in chapter 2), it's easy to understand why Oculus would be enthusiastic about this VR use case as an opportunity to expand the potential market for VR. But is the rapid uptake of VR in these use cases a possible sign that VR has finally identified a problem it can solve?

To understand the success—in terms of funding and adoption—of VR for workplace empathy training, it's worth considering an overview for a moment of the broader context of sexual harassment and diversity and inclusion training in the workplace. In the United States, at least, Title VII of the Civil Rights Act of 1964 made it illegal for employers to discriminate based on race, color, religion, sex, or national origin. Rohini Anand and Mary-Frances Winters write that "this landmark legislation spawned an era of training in the late 1960s and 1970s, largely in response to the barrage of discrimination suits that were filed with the Equal Employment Opportunity Commission (EEOC). If the EEOC or state agencies found 'probable cause' for discrimination, one of the remedies was typically a court-ordered mandate for the organization to train all employees in antidiscriminatory behavior."[44] In 1977, feminist legal scholar Catharine MacKinnon put forward the argument that sexual harassment in the workplace constitutes a form of sex discrimination, which was ultimately upheld by the Supreme Court in 1986. By 1997, 75 percent of companies in the United States had developed mandatory training programs for explaining and reporting sexual harassment, and in 1998 the Supreme Court ruled that harassment training and reporting procedures were sufficient to protect companies from hostile-environment harassment lawsuits.[45] As a result, by 2003, it was reported that companies in the United States were spending $8 billion a year on diversity training.[46]

Yet despite all this training, sexual harassment remains widespread. Based on their meta-analysis of data from 805 companies across thirty-two years,[47] Frank Dobbin and Alexandra Kalev conclude that sexual harassment training and reporting procedures "amount to little more than managerial snake oil."[48] As Nicolás Rivero summarizes in *Quartz*, "It's unclear if any diversity training actually works to transform discriminatory behaviour." Even where diversity training can convince a trainee that discrimination is real, there's little evidence to show that people will act differently based on that new information. In fact, some research shows that men who are more likely to harass women become *more* accepting of harassment behaviors after training.[49] In 2016, the EEOC concluded that "too much of the effort and training to prevent workplace harassment over the last 30 years has been ineffective and focused on simply avoiding legal liability." The EOCC consequently set out new recommendations that employers should "explore new types of training to prevent harassment, including workplace civility and bystander intervention training."[50] It is against this backdrop of a weakening legal shield for existing forms of sexual harassment and diversity and inclusion training, and contemporaneously with the #MeToo movement's phenomenal cultural shift, that VR for (workplace) empathy captured corporate imaginations.

But does it work? Two recent meta-analyses of VR's use training and education found that there is a lack of evidence to support claims that VR works better than traditional classroom lessons,[51] although neither of these studies discussed sexual harassment or diversity and inclusion training. Quoted in an article at Quartz, the author of one of these studies—Lasse Jensen—concluded that one of the key barriers for VR is the difficulty in adapting lessons to specific students: "When instructors can't edit the materials themselves and fit them to the right level for the students . . . it always becomes a bit bad," Jensen said. "This is completely missing in VR because it's so expensive to make one simulation that you can't easily adjust it afterward."[52] A 2022 study by Shannon Rawski, Joshua Foster, and Jeremy Bailenson that is the first to experimentally compare 2D video with a VR video in sexual harassment training had mixed findings. On the one hand, VR increased user intention to deploy informal and nonconfrontational types of bystander responses if they see sexual harassment. On the other, this effect was not present for formal and potentially confrontational responses, and those in the VR condition "explored significantly fewer bystander response options during the practice session compared to the 2D video condition."[53]

Through a more theoretical lens, it's easy to understand why VR isn't a magic bullet for tackling issues as ingrained as sexual harassment or racism in the workplace. To borrow a line from Janet Murray, "Looking down and seeing breasts will not give males sudden insight into the experience of females."[54] VR experiences are not frameless; they are framed by our experiences as a person, leading up to that VR experience. Empathy requires more than just seeing an isolated incident of sexual harassment from the first-person perspective before teleporting safely back to the user's office and comfortable, white, male body. Walking in another person's shoes isn't necessarily enough to go beyond the solipsistic gratitude of "I'm glad I don't experience that" or the egocentric "It's crazy to feel sexually harassed." Although Rawski's study found that participants reported a higher *intention* to intervene—informally, and nonconfrontationally—if they saw sexual harassment in the future (a standard measure for this kind of research), it remains unclear if the necessary imaginative leap took place to generate a real empathetic response that might lead to real and lasting behavioral change.

Irrespective of the lack of evidence, the fantasy of VR for (workplace) empathy is attractive to employers who need to be seen to be doing *something*. The status quo of a PowerPoint presentation was no longer sufficient. For this problem, VR is particularly good. For most users and corporate audiences, VR is innovative and exciting technology. Its use is suitable for press releases; it provides an opportunity to receive commendation for efforts, even if the results do not eventuate. It's also—as most VR start-ups in this space claim—cheaper and more scalable than traditional methods (even if this scalability is at the cost of adaptability and, thus, effectiveness). Tech solutionism like this also has an added benefit, in that it implicitly absolves managers and employers from the harm their prior inaction has caused because they can't be held accountable if they didn't previously have the tools for the job.

What's potentially harmful about this is that VR, with its veneer of authentically real simulations and even greater capacity for data collection and analysis (see chapter 6), inspires an overconfidence in its efficacy that may be getting in the way of actions that may make a real difference in combating harassment and discrimination in the workplace. It's reasonable to conclude that this deployment of the empathy fantasy may be doing more harm than good.

WHO SAYS EMPATHY IS THE SOLUTION ANYWAY?

There's one final point worth discussing around this fantasy of empathy: Even if we discount all these criticisms, if we concede that appropriately designed VR experiences *can* engender an empathy of some sort and that VR's capacity for doing this is greater than, or at least different than, that of traditional media—is more empathy necessarily good?

One of the more viral examples of using VR for empathy is a workplace "empathy training" technology developed by VR start-up Talespin, a US/ Netherlands extended reality (XR)-based learning company founded by Stephen Fromkin and Kyle Jackson in 2015. Talespin was founded with a dual focus on training for the insurance industry via a partnership with Farmers Insurance, and highly real simulations and soft skills training for managers and business leaders. In 2019, Talespin exhibited a demo of its Virtual Human technology, which puts "the user in the shoes of an HR

Figure 3.2   Barry getting fired in Talespin's virtual reality training module (via Talespin).

manager tasked with terminating a fellow employee named 'Barry.'" The press release goes on to describe how "the scenario captures the real stress and emotion typically associated with this situation and presents trainees with common wrongful termination pitfalls such as demonstrating bias. The AI-enabled software provides real-time feedback so a trainee can gain virtual experience that feels real enough to create emotional muscle memory and get real-time guidance on how to empathetically and effectively terminate an employee."

The highly realistic Barry—an older adult in his sixties (see figure 3.2)—uses artificial intelligence to respond to the behavior of the user. If your responses are deemed too aggressive, Barry will express dismay at his chances of finding employment elsewhere at his age. Too soft and Barry might get angry and pound his fists on the table. In interviews, Talespin cofounder Kyle Jackson is quoted as saying, "We have had a number of situations where Barry here, as he starts to get really upset, its emotionally received that way and people get really uncomfortable, their hands start sweating, we've had people actually tear up and cry, we've had people take the headset off and say that is too uncomfortable"—which, Jackson emphasizes, "is a large part of what we're trying to accomplish."[55] Following the negative media attention that this application received (the *LA Times* headline read, "Barry Sobbed as He Begged for His Job.

VR Is Getting Heavy, Man"), Talespin claimed that the Barry demo is just a demo, and no clients have requested to use this software to train employee termination. However, Talespin cofounder Stephen Fromkin still suggests that the Barry demo demonstrates "the emotional resonance of a virtual human for learning."[56]

We can situate Barry, and the broader fantasy of training for soft skills with VR, in the broader history of concepts like emotional intelligence. First popularized in Daniel Goleman's 1995 book by the same name, *emotional intelligence* is often circulated now as a person's emotional quotient (EQ). Goleman argues that emotional intelligence is "as important as IQ for success" in life, but particularly in the modern workplace.[57] VR start-ups specifically call to this concept in pitching their solutions. Vantage Point, for instance, describes how its platform can "train people on EQ-driven skills and soft skills that matter," and Strivr claims its experiences "will help learners practice their emotional intelligence." Mursion even claims that "empathy [is] increasingly identified as the number one leadership skill needed to succeed in today's workplace." The idea of the VR-enabled empathetic manager mirrors the idea and ambition of the emotionally intelligent manager to "transform the effectiveness of organizations" through better management of how employees *feel*.[58] That is, more empathy is good because it will help managers improve profits by helping employees feel better about their working conditions.

Sam Heft-Luthy points out that "in the calls for greater emotional intelligence as the answer to our economic situation, there is a core assumption that the harm to workers under capitalism comes not from a system which denies labor access to the value it creates, but fundamentally from the forms of affect that this system embodies." VR's fantasy of empathy is simply a tool for the managerial class to resolve the affective harm to workers under capitalism, while continuing the underlying exploitation. Heft-Luthy asks, are we simply to accept that a "more empathetic managerial class . . . would provide substantively different outcomes to those we see today?" Or is the fantasy of VR for empathy (again, assuming for one moment that VR can even engender real empathy) just another example of VR's suitability for tech-solutionism, a way for organizations to show material investments in "solving" the affective harms of their workplace in order to avoid having to address structural and underlying issues that cause these harms? As with VR-based diversity and inclusion training, is VR simply being employed as a barrier to real change?

VIRTUAL REALITY FOR GOOD

VR undoubtedly has the potential to be used for good. The purpose of this chapter isn't to dispute this point, but rather to contribute to it. Acknowledging the limitations and risks of VR's capacity to afford presence, to place us in situations that feel *real*, will help unlock the true potential of this emerging medium. To return to the zoo: VR cannot allow us to know the experience of an animal, nor can it perfectly replicate the experience of a genuine, serendipitous encounter with an animal in the wild, but it can enhance our relationship with animals and the natural world, just like seeing an animal in the zoo or attending a zookeeper's talk. Where VR is more practical, and cost-effective, it should be used. While the use of VR in the workplace remains problematic, this doesn't mean that it won't afford new ways of training and addressing bias if implemented properly. At an Oculus Connect panel, Talespin CEO Kyle Jackson discussed how the company found that users would significantly change their behavior when the avatar of the sales target was changed (e.g., from an old white male CEO to a young black female CEO). VR cannot solve bias on its own, but it may become a useful tool, among others, in starting conversations about bias.

The fantasy of VR for empathy reached a fever pitch in 2015–2017, but the notion of the ultimate empathy machine has mostly petered out in the face of widespread criticism. Still, other manifestations of this idea have taken hold, as we see in the use of VR in the workplace and more recently in bullish metaverse rhetoric. Perhaps this highlights how VR's uniqueness can easily be appropriated for new fantasies, recycled and regurgitated in different ways for different audiences. As we've unpacked, the fantasies that emerge, become popularized, and are circulated and recirculated do so because they serve the interests of those circulating them. Questioning the fantasy of empathy, and other manifestations of it, is important because of the potential for harm that VR presents.

VR's fantasy of empathy and more broadly scoped programs like Oculus's VR for Good have another flaw. They mirror the kinds of problematic technological solutionist attitudes present in well-known efforts such as Elon Musk's carbon capture competition and in Web3 discourse. This framing of technology seeks to justify technological development and expansion on the premise that these technologies will ultimately solve society's problems, despite the fact that many of these same technologies contribute to those problems and that other solutions may already exist. Oculus's VR for Good campaign is really just VR for *anything other than gaming.* VR for safety in mines, for heritage, for the emergency room, for action, for understanding, for culture. Anything to expand VR's potential market and justify Meta's fantasy that VR might become as pervasive as the mobile phone (see chapter 4).

So, what can we make of these fantasies? Of VR for empathy? Of VR for good? To conclude, we ask, who do these fantasies serve the most? The fantasy of empathy certainly doesn't serve Sidra, the twelve-year-old girl who was the focus of an eponymous documentary. As VR researcher Kate Clark recently pointed out, there's no information available about what has happened to Sidra after she narrated her story in the Za'atari refugee camp in Jordan. In contrast, Chris Milk's VR/AR company that produced *Clouds over Sidra* went on to release a music fitness VR app called *Supernatural* in 2020, and in October 2021 Meta announced that it was acquiring Milk's company for a reported $400 million.[59]

4

FANTASIES OF ENCLOSURE

In May 2014, flush with cash from a recent initial public offering—the then-highest-valued IPO in the history of American corporations—Facebook CEO Mark Zuckerberg took to his public Facebook page to announce his company's acquisition of Oculus—the gaming-focused VR start-up that had been attracting the attention of the tech press and investors. As he put it: "Mobile is the platform of today, and now we're also getting ready for the platforms of tomorrow . . . Oculus has the chance to create the most social platform ever, and change the way we work, play and communicate."[1]

As we near a decade since Oculus's acquisition, VR is now a central part of the Facebook brand. This was exemplified in October 2021, when the company sought a new identity: the company chose the name *Meta*, a prefix derived from the Greek word for *after* or *beyond*. The name change—a symbolic move (rather than a corporate restructure, à la Google and Alphabet)—reflected what Zuckerberg saw as a distinctive shift in the future of the company's corporate strategy. This was a shift from its family of apps to a focus on the *metaverse*—an imprecisely defined term drawn from science fiction to characterize what the company imagines as a technological stack reliant on new forms of *spatial computing*, a form of HCI that retains and manipulates referents of real objects and spaces (whether material spaces like the built environment, or the organic space of the human body).

With Meta, Zuckerberg proposes a vision that sees his technology empire providing both the software and hardware for a transforming society through its advances and significant investment in VR, AR, wearable tech, and smart home technology—with the company looking to usher in, as Zuckerberg puts it, a "new paradigm for computing and social connection."[2] Yet while Zuckerberg often speaks of VR as a mechanism for delivering something qualitatively new—with the metaverse an effort to move beyond the current moment of computing—things are really just business as usual for Meta. Meta's fantasy is not merely one of enriching social life through technology, but of *enclosure*—of creating a ubiquitous environment owned and operated by Meta, allowing for the unprecedented commodification of more and more aspects of our lives.

OCULUS IMAGINARIES

In chapter 2, we provided an overview of the genesis of Oculus, culminating in Oculus's $2.5 million Kickstarter campaign. The campaign attracted the attention of large Silicon Valley VC companies and, eventually, Facebook—which acquired Palmer Luckey's company for $2 billion. Where some, such as VR pioneer Jaron Lanier, would attribute the Sisyphean struggle (and indeed failure) of VR to achieve widespread use to technical limitations (like the fact that it has been "stuck in a waiting room for Moore's law"[3]), perhaps the more accurate claim is that the VR industry simply lacked capital. A company like Facebook could (and indeed, as we show, would) throw money at VR to push it into existence. For early VR firms—like VPL (which went bankrupt in 1999)[4]—VR needed to turn a profit. But Facebook can simply subsidize its experiments in VR through its high-margin advertising business.

In contrast to its acquisitions of Instagram and WhatsApp, which are both clear efforts at horizontal integration to bolster the company's existing social media arm, it was not immediately clear why Facebook was buying into VR. At this point, VR was a medium that had been a perennial failure as a consumer technology, lacking any real market to sell to. As Zuckerberg stated on his public Facebook account when announcing the acquisition: "Virtual Reality was once the dream of science fiction. But the internet was also once a dream, and so were computers and smartphones. The future is coming, and we have a chance to build it together . . . to start working with the whole team at Oculus to bring this future to the world, and to unlock new worlds for all of us."[5] For Zuckerberg, this would mark the start of a common rhetorical strategy in describing his company's VR investments—one of speaking with the future.

Luckey and Oculus cofounder Brendan Iribe, who had only developed a prototype headset at the time of Facebook's acquisition, saw the Rift headset they were developing as addressing two fundamental problems with consumer VR that had inhibited its adoption. First, high-fidelity VR was reliant upon expensive componentry, limiting its commercial viability. Second, he saw commercial VR as hamstrung by several technical issues: poor image quality (due to low-contrast LCDs), high latency in head tracking (resulting in a discrepancy between movement and the onscreen image, commonly leaving users with a feeling of nausea), and a limited field of vision. Luckey believed that the Rift addressed—simultaneously—the problem of VR's technical impediments and its restrictively high cost. The prototype would ostensibly deliver on VR's long-promised affordance of complete multisensory stimulation, the illusion of spatial coherence created through sensors tracking the movements of the device, inscribed in six-dimensional space. As Luckey put it, reflecting on his naming of the Rift, "I based it on the idea that the HMD [head-mounted display] creates a rift between the real world and the virtual world."[6]

As we showed in chapter 2, for Luckey, the main imagined use case was gaming. Yet Facebook, at least immediately following its acquisition of Oculus, advanced a different vision. Facebook framed Oculus as fitting within its family of apps—tied into the discourses surrounding the family of apps as "social infrastructure"[7] (which, as shown by reliance on Facebook and WhatsApp as primary mediums of communication in some parts of the world, is less rhetorical flourish and more an actuality). From the company's very first Oculus Connect conference—Facebook's VR-focused developer conference[8]—Meta's fantasy is of VR as a form of social computing that will take on a similarly pervasive, infrastructural role in society.

Michael Abrash, a veteran computer programmer of video game and tech companies like id Software, Microsoft, and Valve, was hired in 2014 as Oculus's chief scientist. In his keynote address at Facebook's annual F8 conference,[9] he articulated a fantasy about what such a world might look like. For Abrash, through his career as a programmer for Microsoft Windows NT—the graphical user interface (GUI) that would be extended into subsequent iterations of the now-ubiquitous Windows operating system—computing has long been something that he's understood in infrastructural terms. VR, Abrash tells us, is the next stage in HCI, an inevitable progression along a historical trajectory, from mainframe computers to minicomputers to PCs to GUIs to the web to mobile and, now, to the current moment of VR. To convince the audience of this transformation, he makes the comparison between a 2016 NVIDIA TITAN X graphics card and a 1977 Z80 vector graphic microprocessor—claiming that the former exceeds the latter in computing power by an order of magnitude of a billion. The point, for Abrash, is that transformations in computing cannot be understood in terms of linear growth. Abrash isn't wrong in saying that computers grow orders of magnitude more powerful each decade—but the work it does here is serving as an effective rhetorical technique for shutting down critics of the company's technophilic and hyperbolic claims. The promise of exponential growth of computing power is the guarantor for endless possibility, ignoring the myriad common reasons why people don't actually like or want VR (such as its motion sickness–inducing effects).

In the first minutes of Abrash's address at the inaugural Oculus Connect conference in 2014, he makes comparisons to various (curiously dystopian) works of science-fiction in film (*The Matrix*) and literature (*Snow Crash, Ready Player One*; the latter supposedly became mandatory reading for Oculus employees). To illustrate the transformative potential of VR, Abrash uses a reference to the Wachowski sisters' 1999 sci-fi film, *The Matrix*. While themes of virtuality are central elements to the film, the real focus is on humanity being unknowingly trapped inside a simulated reality, the Matrix, by sentient and hyperintelligent machines. The simulated reality keeps humanity distracted while the machines harvest bioelectricity generated by the human body as an energy source—not too far removed from critiques of Facebook's current surveillance-centered business model, reliant on the harvest of user data, attention, and engagement, and, indeed, a model that the company has more recently suggested it will be adopting in VR.[10]

While VR is framed as new and transformative, Abrash also describes VR's mediatic effects as familiar ones. He suggests that social life has always been mediated in one way or another. A year after the 2014 conference,

Facebook's vision of VR was still speculative and promissory. Abrash, taking the stage at the 2015 Oculus Connect conference, tells the audience that he can't provide them with any evidence by way of a tech demo to substantiate the vision of the immersive future that he and Oculus have promised since the acquisition. Rather, he tries to sell his vision as credible by way of comparison to the auditory-visual perceptual phenomenon known as the *McGurk effect*—where the aural information a person interprets is conditioned by their visual sense data. When there is conflicting sense data—such as the mouth making a certain lip movement to speak and hearing a different sound—the brain interprets the sound differently (the example Abrash gives is of the incongruency of a person making the sound "Fa"—with their top teeth pressed against their bottom lip, while the audio plays the sound "Ba": you still hear the sound "Fa"). "In my opinion, it's impossible to experience the McGurk effect and not believe that the reality you experience is an inference, not a literal reflection of the real world,"[11] Abrash says. He asks the audience to "extrapolate" the McGurk effect demonstration to "full immersion."

The wider point here, for Abrash, is that mediation (whether VR or otherwise) can "trick" consciousness into perceiving one thing or another—manipulating autonomic bodily processes that exceed the capacity for contemplative reflection altogether. The "real world," he suggests, is entirely an illusion, and "our perception of reality is actually just a best guess."[12] It is here that Abrash compares VR, via yet another *Matrix* reference, to taking the "red pill"—that is, the choice offered to the film's protagonist to learn the true reality of the matrix as a simulation (or remain in blissful ignorance by taking the blue pill). For Abrash, Oculus's work was to engineer the right haptic, auditory, and visual inputs that would unlock this new way of perceiving reality.

For Zuckerberg and others, VR is framed not as simply as an extension of some essential human condition of community or sociality. Presenting an interesting twist on how VR is typically imagined as a kind of posthuman extension of human capabilities, VR is instead framed as enhancing, specifically, a Facebook-mediated form of communication (creating feelings of proximity at a distance), one that has become pervasive through the platform's dominance over the last decade. Put differently in 2022 by Zuckerberg in justifying his company's investment in VR: "Most other technology companies are basically out there trying to design new ways for people to use technology; we're out there trying to design technology to create new ways for people to interact with other people."[13] This echoes once again his well-known claim that Facebook is a company that builds social *infrastructure*.

MAKING VR CONCRETE

In the following years, the company sought to make its vision more concrete, first by releasing the Oculus Rift—a commercial version of Luckey's 2012 prototype. The device—much like other comparable devices (such as HTC's VIVE headset)—required a powerful gaming PC to provide computational power for the device. Ultimately, this meant that the device wasn't quite as viable for a broad consumer market as Luckey and Iribe made out because most people didn't own powerful enough PCs. In 2015, Oculus partnered with Samsung to develop the Gear VR headset, a "mobile" VR headset that used the computational power of Samsung's Galaxy phone to act as the device's display processor (instead of a PC). In 2017, Meta released its first stand-alone mobile VR product—the Oculus Go. As a trade-off for portability, the device was limited in its capacity to render high-fidelity graphics or to sense bodily movements with a high degree of accuracy.

While Oculus was becoming more and more concrete as an actually existing suite of products, the ways that Facebook imagined people using the technology were still distant (and, in fact, as we pointed out in chapter 2, the reality was more that the Rift was making people—particularly women—sick). VR was imagined as connecting end users with other end users, enabling intimacy and affectivity through the specific affordances of the combination of Oculus and Facebook's existing social suites. As Oculus's head of product marketing, Meaghan Fitzgerald, notes, VR offers the capacity to make "meaningful connections"[14] with other users, through the integration of Facebook friends (and Messenger functionality), and the ability to view (and create) Facebook content from VR (e.g., viewing and posting to one's Facebook feed using an Oculus device).

As Zuckerberg put it in his Oculus Connect 6 keynote, Facebook's entry into VR represents a pathway to a new kind of "social computing platform,"[15] part of a wider move on the company's part to aim to use computers for "human," "social rituals" facilitated through giving an enhanced feeling of "presence."[16] Once again, Zuckerberg rhymes the future of Oculus with the claim that Facebook builds social infrastructure. Elsewhere, in a presentation by Facebook's then vice president of virtual and augmented reality Andrew Bosworth at Oculus Connect 6, Oculus is imagined as enhancing the social capacity of Facebook through increasing feelings of proximity at a distance.[17] It is perhaps fitting that Bosworth—who now heads Reality Labs (RL), Meta's VR and AR research division —is making this comment about the platform's VR transition, having previously headed the company during another transition from desktop to mobile. Looking to the future, an advertisement for RL plays during Bosworth's talk—signaling the company's ambition to develop mixed reality interfaces, visually augmenting the world with interactive digital interfaces as to make it feel "more immediate, more intuitive, more natural, more human."[18] As Abrash similarly describes it, Oculus offers the capacity for high-bandwidth computing (compared to the "very low-bandwidth" smartphone), framed as letting "us do more of what makes us human, especially socially."[19]

Framing Oculus as a normalized part of everyday life relied on a fantasy about VR as a domesticated technology. It was focused particularly closely on how it could be incorporated into the quotidian activities that we carry out in our daily home lives. We see this for example during Carmack's keynote at the Oculus Connect 5 conference in 2018 (then in the capacity of Oculus's CTO). Carmack suggests that we alter our expectations of the VR medium, particularly around narratives of full-body immersion (invoking the gestural excess at play that is common in many, including Facebook's own, advertisements of VR). As he puts it: "When you have people swinging around wildly and ducking and bending with VR . . . that's not going to be the reality of the way people are using this product . . . it's going to be a niche thing . . . the classic VR thing is the bending, diving, and chucking things . . . it's exciting, but you don't necessarily want to be doing that every day or even every week."[20] Here, Carmack—one of Oculus's most credible game developers–articulates a fantasy about VR occupying a more mundane role: situated in everyday domestic spaces and used for purposes beyond gaming.

Parallel to the significant financial investment in VR software and hardware that Facebook was making at the time, these framings highlight the ways that Facebook was trying to stabilize—in the eyes of the public, and the company's shareholders—decades of competing discourses about what VR could afford and how pervasive it might become. In doing so, the company was also trying to make concrete Facebook's case for being involved in VR, undermining dominant discourses about VR as a gaming technology or as a failed technology that has never taken hold. This is not to say that others have not previously made claims to the social potential of VR, but that Facebook underwent a period of discursively reimagining VR in a way that tapped into Facebook's established reputation as operating the world's most widely used social media platform.

A SOCIAL FANTASY

In 2019, Facebook announced Facebook Horizons (later launched as *Horizon Worlds*)—a virtual environment that the company principally saw as a "social experience,"[21] enabling opportunities for social connection and content creation, similar to other virtual world-building environments such as *Roblox* or *Minecraft*. In a promotional video, a woman is depicted standing in her living room—Oculus Quest headset-adorned—among the domestic rhythms of her partner in the kitchen behind her. Again, the Quest is just another part of quotidian life—for socializing, for play. The woman—through the headset—is depicted as communicating with other users across the world, the device framed as enhancing her social capabilities through increasing feelings of proximity at a distance. VR is framed in terms of a wider turn to what media studies scholars would describe as *digital domesticity*—where the home is extended as a site of technologically mediated "coordination, recreation, socializing and self-expression."[22] In this fantasy, VR becomes just another of the networked, data-driven, sensing *things* that make up the communications infrastructure of the modern home, no more out of place than a smart fridge or digital home assistant device. Zuckerberg has long wanted to emulate companies like WeChat (the Chinese mega-app that has been adopted across almost all aspects of daily life: shopping for clothes, personal finance, communicating with your partner, etc.). It is here that VR is imagined as the medium for sociality and play (and elsewhere, work, as the company has pointed toward with its more recent entry into the software and hardware enterprise market; see also chapter 6).

But what are the emerging issues with this new platform, the product of these fantasies about VR as an everyday, domesticated and pervasive social technology? Distinct from the sanitized futures presented through *Horizon Worlds'* promotional material, reports emerged from beta testers in 2021 that their avatars had been groped by other users. As Tanya Basu reported in 2021,[23] an internal review by Facebook concluded that the beta tester should've used a tool called *Safe Zone* that's part of a suite of safety features built into *Horizon Worlds*. Safe Zone is a protective bubble users can activate when feeling threatened. Within it, no one can touch them, talk to them, or interact in any way until they signal that they would like the Safe Zone lifted. In other words, the onus is on the user rather than Facebook to do anything about the user safety and harm prevention (reflective of the company's wider attitude to platform moderation, as writers like media scholar Tarleton Gillespie have argued[24]). A 2021 *Financial Times* report quoted a memo sent to employees of Meta by the executive leading the push into the metaverse, Andrew Bosworth, in which he said that moderating how users speak and behave "at any meaningful scale is practically impossible."[25] As he put it elsewhere, "We can't record everything that happens in VR indefinitely—it would be a violation of people's privacy."[26] The specter of privacy—here, presented in a highly individualized sense (the threat of Facebook watching *you*)—is mobilized to absolve Facebook of its own failure to responsibly moderate the behavior of its users.

Amid claims that Facebook more broadly has been grossly negligent in moderating content on its platform—in recent years, particularly mis- and disinformation—there is concern about how the company might prevent its virtual reality from coming to exist as another alternate reality for groups like QAnon. As Buzzfeed journalist Emily Baker-White found, Facebook was slow to process take-down requests for a *Horizon Worlds* room in which the journalists posted words and phrases that they believed would trigger Facebook's system for monitoring community standards violations (such as COVID-19 conspiracy theories). What became clear is that the company was relying on blocks, mutes, and reports to notify it of community standards violations. The company's approach to content moderation in VR as it currently exists is at odds with its vision of VR as a medium for social life.

It is here, where Facebook's highly sanitized vision of VR rubs up against a milieu of misogynistic internet culture, that Abrash's invocation of the red pill (as envisioned by the Wachowski sisters) takes on a very different meaning, more in line with its usage in contemporary internet parlance: choosing to open one's eyes to the oppression of largely cisgender white men in light of what is pejoratively called *identity politics*. Where the technology is conceived of by Facebook as representing freedom and autonomy, frictionless embodied communication, it is clear that this is not the case for the groups against whom the technology is weaponized.

THE FANTASY OF A VR ECOSYSTEM

Facilitating the shift toward more everyday uses, in 2018 Facebook announced the Oculus Quest—a new "all in one" mobile VR headset, far more powerful and capable than the Oculus Go. In contrast to the Rift line of headsets—which required external sensors and for the device to be connected with a sufficiently high-performance computer—the Quest has all componentry built in: a "stand-alone" system. As Facebook's technical documentation outlines, the Quest's technological stack is reliant on a form of computer vision algorithm known as visual-inertial simultaneous localization and mapping (SLAM). Put simply, SLAM is a computational method of constructing a digital map of the environment the device is located within. SLAM enables the device to know where it is, relative to where it was, thus enabling it to properly calibrate motion.[27]

As mobile media scholars Michael Saker and Jordan Frith put it, the Quest represents a "coextensive space," a "way of understanding the developing relationship between the physical, digital and concrete reality that is being enacted by current VR systems"—one that is "forging an altered relationship between the physical, the digital and concrete space, through the mediated inclusion of concrete reality."[28] As Zuckerberg stated at Oculus Connect 5 when announcing the device, mobile, stand-alone headsets—as distinct from the wired headsets on offer from market competitors—will be necessary in scaling VR, part of the company's goal of reaching one billion VR users.[29] In a *WIRED* magazine interview with journalist Steven Levy, Zuckerberg notes that VR represents an opportunity for Meta to scale a full stack of technologies—both hardware and software. Where previous attempts to do so have failed (such as a failed internal project in which Meta tried to develop its own phone and operating system, but found it was "too late to topple the dug-in lords of mobile"[30]), Meta has the first-mover advantage in the VR market.

The fantasy of an XR "ecosystem"—a systematic effort to grow the company's wider embodied computing offerings, rather than just focusing on VR—emerged in 2018, leading up to the release of the Quest. For example, it was here that Facebook announced that its Oculus headset would interface with applications like Facebook and Instagram. To enable this vision, in 2020, Facebook announced that Oculus accounts opened in January 2023 would require connected Facebook accounts, a move that provoked widespread anger among Oculus's early supporters and an investigation by Germany's national competition regulator.[31]

To achieve the scale and integration with everyday life that Zuckerberg desires, the growing XR ecosystem requires not only end users, but also third-party complementors. These include both software developers and businesses, drawing directly from the company's playbook with Facebook: it grew out its ad services through partnerships with advertising technology companies, with these third parties giving Facebook much of the data used for their targeting algorithms. Developer partnerships were a common form of corporate partnership across Reality Labs. Encompassing AR and VR and promoted by (but preexisting the formation of) Reality Labs, this approach included partnership programs like Oculus Launchpad (a development incubator), which centered on providing resources and promotions for emerging VR development (with an emphasis on groups typically underrepresented in VR development).[32] Subsequently, in 2021 an XR Hackathon was announced by Meta: "We're excited to invite global developers and creators to build immersive XR solutions for the chance to win a total of $700,000."[33] The XR Hackathon—more focused on Reality Labs' ambitions than the broader Facebook Hackathon scheme—centers on numerous key areas of research and development, such as "Hand and Body Tracking Performance AR Effects, and Voice and Hand VR Experiences."[34] Other programs include the Creator Accelerator

Program, a creator incubator that has a focus on teams developing content for the company's metaverse software, *Horizon Worlds.*

To facilitate developer engagement, Facebook sought to make Oculus a modular platform through its Oculus software development kit (SDK), a suite of tools for the development of software across Oculus's devices. For example, one update to the SDK facilitated VR software development with the popular Unity and Unreal game development engines—drawing into the ecosystem developers working with these common software tools. Elsewhere, the SDK enables compatibility with OpenXR—an XR development standard that provides an API to allow developers to work across a variety of AR and VR devices (as opposed to working with numerous proprietary APIs, locking developers into developing for particular devices). Theoretically, such a development standard would be beneficial for the medium of XR development in general, rather than Meta specifically. There is a sentiment of interoperability that Zuckerberg would go on to echo in 2021, announcing that "the metaverse will not be created by one company."[35] Openness is a means through which Meta signals good intentions of bolstering the nascent XR ecosystem. But at the same time, it is also the means through which Reality Labs is structuring and deriving value from multisided interactions—specifically, by encouraging development on Reality Labs' platform. For the company that currently dominates the VR market with over two-thirds of the market share, this gesture toward decentralizing its ecosystem means little, in effect further producing economically and structurally centralized outcomes for Meta.

The role of these complementor partnerships, as we have theorized elsewhere,[36] can be understood in social theorist and geographer Clive Barnett's terms as *convening*, "calling out others, attracting their attention," requiring an "active response":[37] here, usage or participation. Through these kinds of developer initiatives—whether through more formal partnerships or community hackathons—convening seeks to achieve two goals: the outward expansion of Oculus through the creation of a developer ecosystem, and the creation of content to attract or maintain end users. The aim, as Zuckerberg suggests, is Meta's AR and VR being "completely ubiquitous in killer apps."[38] As some critics have pointed out, Meta's insistence on growing its Oculus app store (and a VR ecosystem reliant on it) is less of a regulatory headache than running a social media platform—where it has drawn the ire of regulators globally. To meet this need, Oculus announced in 2021[39] that in addition to its Oculus app store (its main digital distribution platform, from which it takes a 30 percent cut of sales), it would allow distribution of experimental and in-progress works, subject to a less strict process of review than apps in the app store, through the App Lab platform (from which the company takes a lesser 15 percent cut of sales): a less restrictive pipeline for development from which Meta profits.

The growth of Oculus's developer ecosystem has not been unproblematic. One point of criticism that emerges from VR communities and developers centers on Facebook's alleged co-opting of third-party software published on its app store.[40] Internet studies scholar William Partin notes that this is a common modus operandi of platform companies—extending their tentacular grasp over not only users, but platform complementors. As he puts it, "The technical architecture of platforms evolves through the exploitation of power asymmetries between platform owners and dependents."[41] While Meta's Oculus developer programs purport openness and opportunity for developers, they are underlain by insidious anticompetitive practices of platform capture (drawing the ire of regulatory bodies in the United States, such as the Federal Trade Commission and the Department of Justice).[42]

Other research would suggest that Oculus's attempts to grow a third-party ecosystem are problematic in terms of data and privacy. Computer scientist Rahmadi Trimananda and colleagues compared network traffic and audited privacy policies of apps published on the Oculus app or Sidequest store, finding that "approximately 70% of OVR [Oculus Virtual Reality] data flows were not properly disclosed" and that "69% of data flows have purposes unrelated to the core [software] functionality."[43] In line with a range of interdisciplinary academic critics and privacy advocates, the authors "found evidence of apps collecting data types that are unique to VR, including biometric-related data types." As our own audit of Oculus's privacy policies (from 2014 to 2020) has found, while the company does specify data types and uses, this is done in broad-brush terms—so much so that they are rendered effectively useless in terms of understanding what the company may use collected data for.[44]

THE MONEY CANNON

The social fantasy that Facebook proposes is one that the company sees as requiring investments in research and development capability. Let's jump back to 2014. Following the Oculus acquisition, the company launched an internal research division called Oculus Research—headed by Abrash, then newly hired. As Abrash puts it, in the announcements of Oculus Research during his 2014 Oculus Connect talk and his 2015 F8 keynote address, the division represented the "first complete, well-funded VR research team in close to twenty years."[45] In one promotional video, Oculus research is described by Abrash as "set up around the whole idea of rapid iteration; we know a lot of it will fail, but some of it will succeed and change the world."[46]

The contributions he suggested were going to extend far beyond VR. Abrash compares the impact of Oculus Research to numerous key research centers in the field of HCI, from the Augmentation Research Center at Stanford Institute (and the work of Doug Engelbart on the GUI in particular) and Xerox PARC (another research center instrumental in the development of many innovations in modern computing, including the computer mouse). Notably, PARC, while producing technological and conceptual innovations that would shape the trajectory of computing, was famously unprofitable for Xerox—and it was ultimately other companies who benefited from PARC's inventions (such as Apple, which drew on PARC's innovations in developing the computer mouse). Yet Abrash suggests that Oculus Research will succeed where PARC failed, particularly in that the goals of Facebook as a business are aligned with those of the lab: both embody a strident commitment to VR development (and now spatial and embodied computing more generally)—where PARC's goals, Abrash argues, were misaligned with Xerox's business model of making and selling printers and photocopiers. The lab itself was involved in the development of numerous technological advances in VR instantiated in various iterations of Oculus devices, such as hand tracking and odometry (i.e., the computational process through which a device, such as a Quest, is able to determine changes in its position). Oculus Research hosted research scientists, developing in-house research, and was engaging with academic disciplines such as computer science and HCI. Oculus Research, along with conducting research internally, was engaged with academia, publishing with the ACM Special Interest Group on Computer-Human Interaction (SIGCHI) and the ACM Special Interest Group on Computer Graphics and Interactive Techniques (SIGGRAPH).

Fast-forward to roughly where we left off in the previous section. In May 2018, Facebook shuttered Oculus Research, founding Facebook Reality Labs in its place. Reality Labs is Facebook's mixed reality and spatial computing "research and development group,"[47] comprised of twelve individual labs located across the world, each of which focuses on specific areas of spatial computing research and development (e.g., the London group focuses on Spark AR, while the Zurich group focuses on computer vision—such as that used in the Quest's odometry stack).[48] In 2020, the scope of Reality Labs was described as covering three currently existing technological projects—Oculus (VR), Spark AR (mobile AR), and Portal (smart home assistant technology)—as well as more speculative research and development projects (largely centered on wearable AR).

As Facebook first revealed in a 2021 earnings report, Reality Labs has been a significant source of expenditure—spending upward of $4 billion a quarter since 2019. As we noted in the introduction to this book, to give a sense of the scale of this investment, this is approximately thirty-one times the size of the budget allocated to taxpayer-funded research grants in Australia, and roughly equivalent to the entire research and development expenditure by businesses in the country. As one ex-Oculus employee told us, the amount of money Meta was prepared to spend on VR and AR projects at Reality Labs earned the organization the internal nickname of the *money cannon*. While this spend isn't *just* on VR (with money going into some of the AI-centered mixed reality projects), Facebook provided little insight to the public (and the company's investors, to their frustration) into where exactly these funds were allocated.

Through Reality Labs, the company has systematically sought to expand its VR capabilities, buying out a range of VR-related firms between 2018 and the present.[49] Relative to its other big tech peers, Facebook's corporate strategy has not historically focused on acquisitions of other firms, but since 2018 alone it has acquired fourteen companies, including game developers and VR software developers. For example, in 2019, Meta purchased Beat Games—developer of *Beat Saber*—which the company revealed was the "highest earning app on Meta's Quest Store to date by revenue," with the game's 2021 revenue "greater than the next five highest-grossing apps combined."[50] In 2022, Meta made its most expensive acquisition—its $400 million acquisition of Within (despite an intervention by the FTC to block it), a VR fitness company—with Zuckerberg in 2022 describing his ambition to create a subscription-based service akin to Peloton (the exercise-equipment-as-a-service company that was one the worst-performing tech stocks that same year).[51]

Beyond its consolidation of the VR software market, Facebook was also increasingly buying out firms that developed some form of mapping or machine vision technology. These generally fell into two subcategories: AI companies interested in capturing and/or rendering data about physical environments, and AI companies that capture and parse body activity. In 2019, Reality Labs acquired CTRL-labs—a neural interface start-up. At the time of acquisition, CTRL-labs was developing wristbands that translate neuromuscular signals into machine-interpretable commands— which in 2021 form a key part of Facebook's narrative about the future of Reality Labs' computing interfaces,[52] with CTRL-labs now forming one of Meta's twelve Reality Labs divisions (its New York lab, focused on neurotech interfaces). As Reality Labs highlights in a demonstration of internal research, CTRL-labs has played a central role in the development of brain-computer interfaces that the company seeks to pair with VR and AR hardware. In 2020, Reality Labs acquired Lemnis Technologies, a headset manufacturer with a specific focus on eye-tracking technology. Eye tracking has since become a focus for Reality Labs in what it named *Project Cambria*—the development codename for what would eventually be launched as the Meta Quest Pro, a VR system with improved sensor abilities.

Beyond interfaces for sensing the body, Reality Labs' other AI acquisitions were to do with tracking and graphically rendering the physical environment. On the stage of its 2019 Oculus Connect conference, Reality Labs showcased its vision for *LiveMaps*—a project for creating a digital map of the world to render it augmentable through technologies like smart glasses (which the company would later develop into more expansive AI projects for tracking and rendering spatial data; specifically, data about the built environment, such as the urban environment or one's living room). We see this in specific internal projects like Project Aria in 2020 (which created 3D scans of the environment that could be modeled in AR and VR headsets) and Ego4D in 2021 (which used AI to analyze 2D, first-person video recorded through prototype AR headsets). Coinciding with this vision, Reality Labs acquired numerous AI companies that develop software for mapping and computational modeling of physical space. In 2019, Reality Labs acquired GrokStyle, an AI company centered on mapping spaces in AR (with a particular focus on retail, such as placing digital representations of furniture). A further range of acquisitions relate to Reality Labs' announcements from 2019 to 2021 concerning augmented reality and digital mapping. In 2020, Facebook acquired Mapillary—a mapping company for crowdsourcing maps based on image data—and integrated it into Reality Labs. Likewise, that same year, the company acquired Scape Technologies, developer of a 3D mapping and scene reconstruction technology. At time of acquisition, Scape Technologies was building a visual engine that allows camera devices to understand their environment using computer vision. This builds on the use of similar techniques that Facebook employs in the odometry stack of its Quest console.[53]

The work at Reality Labs—beyond an investment in first-party software, and research and development capacity—was an effort at creating the optimal social and political conditions for the company's VR ambitions. Over the last decade, Meta and its family of apps—namely, Facebook and Instagram—have been subjects of intense scrutiny (sowing the seeds of distrust for the company as they enter into the VR space). It is also the case that regulators and policymakers are beginning to raise concern over Meta's entry into the VR market (raising issues to do with competition, privacy, and consumer safety[54]). As Meta grows louder in proclaiming VR as the next iteration of computing, so do its critics.

To mitigate the effects of the considerable distrust toward the company as a custodian for VR and other spatial computing technologies, Reality Labs entered into a number of partnerships with academics and policymakers—a dynamic that can be described as *capture*,[55] a form of control exercised by powerful corporate institutions over other institutions (even very powerful ones, such as academia and government)—creating a channel between these (noncorporate) spaces and corporate interest.

One major example emerged concurrently with Meta's September 2020 announcement of Project Aria, an internal research and development program for a pair of AR glasses. The company imagines that these glasses will augment physical space with contextually specific information using AI. The company published a number of best practice principles for its development of AR. The principles generally sought to address expectations of risk that were immediately raised with Project Aria, and that thus likely lie ahead in the future of AR. They are "never surprise people," "provide controls that matter," "consider everyone," and "put people first."[56] As Reality Labs notes, these principles were developed in consultation with "external experts," yet the company does not state who these individuals and groups are. As critics like Applin and Flick[57] argue, Reality Labs' responsible innovation (RI) statements amount to little more than an attempt at preempting critique of Project Aria, particularly considering previous internal experiments (such as the infamous emotional contagion study).[58]

Coinciding with its RI policy for Reality Labs, Facebook Research—the company's academic research division—broadened the scope of its research focus areas to include Reality Labs–related projects covering AR and VR. For instance, in 2020 and 2021, as part of Facebook Research's academic grant scheme—in which it attempts to solicit research from academics external to Facebook—AR and VR became priority areas, the focus of two funding rounds (of $75,000 grants). These focused on "explorations of trust in AR, VR, and smart devices" in 2021—emphasizing "security, privacy, integrity, and ethics," and "responsible innovation in AR/ VR" in 2020, likely a response to critique of Facebook's VR emerging at the time by civil society and privacy advocacy groups and by regulatory bodies in Germany throughout 2020.[59]

Throughout 2020, Reality Labs' partnerships and attempts to drum up goodwill were increasingly focused on the policy space—likely as a response to growing critiques of its practices within Reality Labs specifically, and across the company more broadly. Reality Labs' partnerships since 2020 have been centered on what Meta refers to as *responsible innovation*. RI is a general approach to technological research and development that seeks to embed social benefit and moral responsibility, centered on areas like "anticipation, reflexivity, inclusion and responsiveness."[60] As critics see it,[61] RI is a common way for technology companies to offset negative perceptions of societal harm. For companies like Facebook, RI operates as what Thao Phan and colleagues refer to as *virtue capital*—by which perceptions of goodwill and trust are necessary industrial inputs to maintain a system of production.[62]

In September 2021, the company announced a $50 million investment focused on regulation and policy—an effort to engage lawmakers and academics. The investment, the XR Programs and Research Fund, came shortly after allegations of impropriety in its platform governance by company whistleblowers. The fund sought to develop research, regulation, and policy for the metaverse, centered on areas of risk such as privacy, safety and integrity, and equity and inclusion. The fund was announced by CTO Andrew Bosworth and VP of global affairs Nick Clegg. Clegg—previously the UK deputy prime minister in David Cameron's Conservative-Liberal coalition—was hired by Facebook in 2018, effectively as a well-connected lobbyist and spokesperson. Much like his function in the Cameron administration—where his appointment created a veneer of social progressiveness—Clegg's appointment by Zuckerberg was another strategic move to appease regulators by showing Meta's willingness to play ball. For example, in 2020, when the company was under intense scrutiny for its algorithms amplifying hate speech and inflammatory posts (facilitating the genocide of Rohingya Muslims in Myanmar), it was Clegg who Facebook would call out to face the press or government regulators and deflect blame. As Clegg writes in a 7,900-word blog post, published in May 2022, the company is "committed to building [the metaverse] in a responsible way"—through efforts such as the XR Programs and Research Fund. "The metaverse is coming, one way or another," he wrote.[63] Here, Clegg divines a *telos*, or final cause, of the internet. Fate is already written, through the investments of Meta and other tech firms in the metaverse. And because the metaverse is coming, it must be regulated, argues Clegg. As he puts it, "The metaverse will bring with it huge potential for social and economic progress. And it will bring risks and challenges, many of which can be anticipated. Our hope is that the lessons of previous technological advances can be learned and that the rules, standards and norms that will govern the metaverse can be developed in tandem with the development of the technologies themselves."[64]

Through Reality Labs, Meta develops capabilities in growing the underlying technical infrastructures for its VR and spatial computing offerings. It has sought to consolidate competing firms through acquisitions and attracted third-party complementors into its ecosystem. Recognizing the important role of social and cultural forces—such as public and policymaker perceptions of technology—Reality Labs has also actively sought to capture critics and regulators, through efforts to launder its business practices and internal research and development, drumming up goodwill at a time when the company is under more scrutiny than ever.

CUSTODIANS OF THE METAVERSE

Through Meta, Zuckerberg proposes a vision that sees his technology empire providing both the software and hardware layers to support a digital society, through its advances and significant investment in VR, AR, wearable tech, and smart home technology. As Meta envisions it, how we engage with (mediated) space has implications for user subjectivity and agency, with key applications to come for "how we work, play and connect with one another."[65] As it was framed at the company's Connect conference in 2021, this shift toward spatial computing is part of the company's push toward what it refers to as a *metaverse*—that is, what Zuckerberg describes as an "embodied internet"[66] that is at once based on an "interoperable" network of computing platforms while also being centralized through a system dependent on Meta's own hardware and software.

The metaverse—despite extensive coverage in the news and tech trades—is not an altogether new idea. Beyond the term's basis in Neal Stephenson's 1992 dystopian sci-fi novel *Snow Crash*, in which the metaverse describes a future where digital and physical worlds blur and people live their lives through virtual and augmented reality-driven interfaces, the term has seen a resurgence in recent years in technology venture capital circles—from stakeholders who exert considerable imaginative force in shaping the collective commitments of institutional investors, as well as industrial entrepreneurship and innovation. In 2019, Matthew Ball—a venture capitalist and author of the influential book *The Metaverse: And How It Will Revolutionize Everything*—wrote an essay locating the concept in interoperable digital worlds (in the context of the growing popular-ity of the game *Fortnite*).[67] Technology VC giant Andreessen Horowitz—a firm that was one of Facebook's earliest investors, and with Marc Andrees-sen sitting on Facebook's advisory board—has likewise been pushing the concept since at least 2020.[68] Most of these are vague visions of a new kind of internet (often drawing on ideas from previous hype cycles for virtual life, such as virtual work). Tech companies—such as Microsoft, Mozilla, NVIDIA, and Samsung—have begun invoking the term, which has become a catchall (and thus not particularly useful) descriptor for any kind of virtual environment. The term operates as a floating signifier—a term with no common, agreed-upon meaning, a kind of definitional pro-miscuity allowing it to appeal to a range of different constituencies and their interests, something that provides it an alibi for extending into dis-parate areas of life. In recent years, the term is often appearing alongside the equally as nebulous *Web 3*—the so-called next iteration of the inter-net, one ostensibly decentralized (in contrast to the walled gardens of Web 2.0) through blockchain technology.[69]

While there are competing ideas about what a metaverse is, it is Zuckerberg's vision of a so-called embodied internet that has garnered the most attention, becoming almost synonymous with the term *meta-verse*. Zuckerberg sees the metaverse as the "next" iteration of networked computing—a vision that he constantly hedges with the statement that these futures are still very distant. In his address to investors during the company's 2022 first-quarter earnings presentation, he notes that he's prepared to spend tens of billions of dollars over the next decade, "lay-ing the groundwork for what I expect to be a very exciting 2030s when this is established as the primary computing platform" (even prepared to "trade off against shorter term financial goals").[70] As Zuckerberg notes in an appearance on the stage of the South by Southwest (SXSW) festival in 2022, "Just believing in things and having a very strong conviction is one of your most powerful tools."[71] This is a point he makes with reference to detractors of the original Facebook website and the company's highly successful pivot to mobile. "If we succeed at this it will be because we care more about that problem [of VR as a medium for social connectiv-ity] and believe in it more deeply than all the other folks, who maybe have decades more experience doing this kind of thing than we do."[72]

Zuckerberg essentially believes that he can manifest this shift in computing that he describes. You might dismiss this kind of statement as the kind of platitudinous (if not deluded) motivational drivel that one might hear from a self-help guru—if it were not being made by the CEO of one of the biggest tech companies in the world. As Neal Stephenson remarks, he was just "making shit up"[73] when he coined the concept of the metaverse in *Snow Crash*. Perhaps the same is true for Zuckerberg.

Facebook's announcement of its own metaverse has not come without criticism and derision. It's juvenile and far-fetched[74]—so much so that taking it seriously simply lends credibility to Facebook's incredulous statements.[75] It's a move of desperation to distract from claims of Facebook's social harm.[76] It's an effort to diversify its business model in light of threats to its advertising service after restrictions on mobile user targeting by Apple and Google.[77] It "simply looks like shit"[78] (as one commentator puts it, in response to the low-quality graphics of *Horizon Worlds*).

Yet we shouldn't be too quick to dismiss the company's real, systematic efforts to grow its spatial computing efforts—things that largely lie out of sight from public reporting. One development has been Facebook's Ego4D project—an effort to train the assistant AI that would be used in its AR and VR smart glasses and future smart home assistant AI such that they can provide contextually specific information (directional prompts in navigation, information about location, etc.). This would see a seven-hundred-user cohort (of Meta employees and contractors) sent into the world wearing prototype glasses to train image-recognition algorithms. A 2020 video, mocking up what the future of the technology might look like, demonstrates the value of this data: we are presented with a fantasy in which a user walks through cities, their vision annotated with contextually specific information about the environment (such as information about a restaurant one user passes by).[79] As Sally Applin and Catherine Flick argue, this fantasy in which spatial computing extends from the home to the urban center is a mechanism through which Meta can "follow us more deeply into the Commons."[80]

The motivations behind this fantasy are not hard to surmise when we think about how Meta makes its money: intermediating the buying and selling of ads. As tech critic Ben Tarnoff writes, the privatized, ad-subsidized internet (and companies like Meta) can productively be compared to a shopping mall—"corporate enclosures with wide range of interactions transpiring inside of them"—one where the platforms act as rentiers, taking a cut of the action.[81] The potential to layer immersive virtual and physical spaces with digital ads takes this to the extreme. While Meta frames Project Aria with knowledge and convenience, it is easy to think of visions of similar futures depicted in artworks like Keiichi Matsuda's 2016 video "Hyper-Reality,"[82] where mixed reality interfaces inundate the user with a deluge of personally and contextually targeted advertising; new (and potent, due to their appearance in our field of vision) forms of algorithmic targeting that lie outside of, yet have the potential to shape, thought and action.

Documents and comments from Meta reveal that ad monetization is an area the company is actively exploring. As documents, emerging from the 2021 Facebook Files leaked by whistle-blower Frances Haugen, would suggest, the company's ambitions for a mixed reality–enabled metaverse are motivated by the monetization of more and more aspects of daily life through the medium of all-day, embodied computing, promising to "generate significantly more ARPU [average revenue per user] than other social graphs."[83] More recently, in an interview with the Verge website, Meta's vice president of augmented reality, Alex Himel, notes: "We should be able to run a very good ads business. . . . I think it's easy to imagine how ads would show up in space when you have AR glasses on. Our ability to track conversions, which is where there has been a lot of focus as a company, should also be close to 100 percent. . . . If we're hitting anything near projections, it will be a tremendous business . . . a business unlike anything we've seen on mobile phones before."[84]

As we finish writing this book in March 2023, it remains to be seen whether Meta will be able to make good on the promises of the metaverse. Notably, in early 2022, the company revealed in its financials just how expensive its highly speculative Reality Labs venture was: the company had been hemorrhaging roughly $4 billion a quarter on Reality Labs alone. The day of this announcement saw the company's stock price plummet. Over $200 billion was wiped from the company's market valuation—the largest single-day drop in history. Although we would be remiss to attribute this drop solely to the company's supremely expensive (and highly speculative) Reality Labs spending (the company also suffered blows to its ad business due to Apple introducing data-tracking restrictions on iPhones), it was a revelation that saw the company derided by tech critics and heavily criticized by shareholders. Some have called for Zuckerberg to be ousted, but the company's shareholder structure preserves his control of Meta regardless of his fixation on the metaverse. Whatever debate there is about the intellectual merit of terms like *techno-feudalism*, Zuckerberg is well and truly the ruler of Meta's platforms.

A bigger problem lies in wider macroeconomic shifts—namely, rising interest rates. Tech companies have enjoyed over a decade of near-zero interest rates, meaning that they could issue (cheap) debt to finance projects that they believed would pay off in the long term (i.e., where return on invested capital would exceed costs paid for its debt). The sheer volumes of capital available to big tech meant that firms could, essentially, will things into existence by throwing enough money at them. No longer able to take out cheap debt to fund its massive Reality Labs–related capital expenditure, it is unclear whether Meta can finance this massively expensive venture on cash alone. Maybe the money cannon is finally out of gunpowder.

In 2022, Meta ran an ad during the Super Bowl promoting its Quest 2 headset. The commercial features an animatronic dog, made obsolete after Questy's—a Chuck E. Cheese–style venue—shuts down. The dog is discarded, only to be rescued from a garbage compactor and (for reasons unexplained) outfitted with a Quest headset, allowing it to reunite with the other animatronics from Questy's in the virtual world. While this moment was likely meant to read as a moment of triumph—of overcoming adversity—in a rather dystopian way, it implies that the real world is bleak and that it is virtual worlds that are the only source of joy. As one YouTube commentor sardonically puts it in response to Meta's video, "We destroyed your world, so you could come play in ours" (yet it's unlikely that this will be much better, if Facebook's current track record is anything to go by). On its face, this seems a curious pitch for the metaverse. But perhaps it perfectly encapsulates the fantasy that Meta has articulated for VR from the very beginning—the total enclosure of everyday life.

5

FANTASIES OF VIOLENCE

In March 2018, Stephon Clark, a twenty-two-year-old Black man, was shot and killed by two officers of the Sacramento Police Department. Police claim that the shooting occurred because the mobile phone he was holding was misidentified as a gun. Clark is far from the last to be killed by police in the United States. In 2020, with a wave of uprisings across the nation spurred by the murder of George Floyd by the Minneapolis Police Department, Sacramento police responded to the killing of Clark (and others nationwide) by announcing that it would require its officers to participate in VR-based training simulations, supposedly equipping them with skills in "anti-bias," "critical decision making," and "de-escalation."[1]

The adoption of VR training simulators by police in Sacramento mobilizes a fantasy of simulation that is more than centered on the haptic, auditory, and visual elements of VR, and the suggestion that VR can deliver higher-fidelity renderings of the world than other forms of computation, a fantasy that appears rather neutral. On the contrary, simulation is mobilized here to rehabilitate the image of policing at a time when many across the world are coming to question its moral legitimacy. VR's simulational affordances are put to work in legitimating an institutionally racist and deeply authoritarian organization. The fantasy becomes an alibi for the expansion of violence that constitutes policing.

As feminist philosopher of science and technology Donna Haraway argues, ways of knowing and thinking about the world bear ideological marks. Ostensibly "objective" models of the world often enact "militarism, capitalism, colonialism, and male supremacy."[2] Although Haraway was writing of the production of scientific models of knowledge, in this chapter we take this as a prompt for thinking about the computationally rendered models of real-world phenomena in VR simulation. VR simulations are made by people (carrying their own logics, ideologies, and biases) to achieve certain ends (benefiting some, but not others). To riff off political theorist of technology Langdon Winner—and his maxim that technologies have politics[3]—we can say that VR has politics. In this chapter, we argue that VR's fantasy of simulation cannot be disentangled from a reality of violence—focusing on violence exerted by the police and military, institutions of the state with historic and current-day relations to VR.

Violence is a term with broad breadth of meaning across law, philosophy, and political theory. Violence can be direct (such as through acts or threats of applying physical force, coercing people to act in a certain way). But violence is also indirect (or "cultural") in the way that "culture, the symbolic sphere of our existence . . . can be used to justify or legitimize direct or structural violence"[4]—such as the aforementioned example of VR as alibi for policing, an effort to absolve the institution of harms inflicted by individual officers.

VR, as we show in this chapter, has long been a technology of violence—emerging in the mid-twentieth century to support the interests of postwar military technoscience and, later, national defense and intelligence operations during the Cold War. VR was used as a tool for training users of weapons systems, as a technology to advance wider military technoscientific developments, and a means for the expansion of US hegemony. But it hasn't stopped there. We aim to establish a more current trajectory in which the militarized legacy of VR has filtered into the imagination, development, and application of specific technologies of VR simulation today.

A HISTORY OF VIOLENCE

The simulational fantasies of VR have their basis in the militarized history of computing in the mid-twentieth century, particularly during (and following) the Second World War. As media historian Bernard Dionysius Geoghegan notes, wartime computing—borne out of projects like cybernetics and information theory—sought to establish "stable ratios between bodies, machines, and space,"[5] convening the modes of human attention necessary for aerial defense or ballistics operations. Computers—in other words—addressed a cognitive deficit in humans, limited in their capacity to accurately interpret information at speed.

It was Ivan Sutherland—an American computer scientist—who, along with a team of doctoral students at Harvard, would develop the first proto-VR application in the 1960s—known as the *ultimate display*. Sutherland got his start designing the interactive graphical computer-aided design system Sketchpad while working on his doctorate at MIT (which would go on influence the broader field of GUIs). While Sketchpad was not designed with military applications in mind, the system was created in a facility dedicated to developing and expanding the US defense system at the end of the Second World War, under the supervision of Claude Shannon (who had previously worked on wartime fire-control systems and cryptography with the National Defense Research Committee).[6] Following completion of his doctorate, Sutherland took a job at the National Security Agency, before taking on a position in the US Army as a first lieutenant, in which he headed the Information Processing Techniques Office at the Defense Advanced Research Projects Agency (DARPA) from 1964 to 1965. The Information Processing Techniques Office was opened in 1961 for command-and-control research, borne out of funding for an aerial defense computer program that would run as a backup to the Semi-Automatic Ground Environment (SAGE)—the United States' primary aerial defense computer network. In practice, this office (under the stewardship of its first head, J. C. R. Licklider, and later under Sutherland) centered on research into interactive computing (as well as ARPANET, the computing network which would become the technical foundation for the internet).

It was after his tenure as DARPA head that Sutherland, now working at Harvard, along with his doctoral students Bob Sproull, Quintin Foster, and Danny Cohen, developed the ultimate display—a computer display with a head-position sensor, which meant that the display could change based on movements of the head. This system is often described as the progenitor to VR (see figures 5.1 and 5.2).[7] The impetus (and much of the funding) for the project came from the American aerospace company Bell Helicopter—which sought a solution to the problem of landing helicopters in narrow clearings at night (the reason *why* isn't entirely

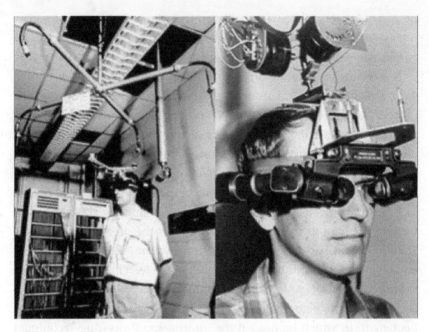

Figure 5.1 Ivan Sutherland's Sword of Damocles head-mounted display, widely considered the first VR headset. This system utilized mechanical pivots to provide the viewer with six degrees of freedom in the head mount for an immersive viewing experience.

clear, although Bell at the time was being contracted by the Pentagon to build stealth helicopters for nighttime reconnaissance missions into North Vietnam). As night vision equipment was bulky, Bell's plan was to mount it on the bottom of the helicopter. Bell experimented with placing infrared cameras underneath a helicopter and attached the cameras to pilots' head equipment (such that when the pilot moved their head, the camera underneath the helicopter would also move), allowing the pilot to see the ground underneath the helicopter (this was ineffectual, as Sutherland would later reveal, operating more as an "attention focuser," one that "defined a set of problems that motivated people for a number of years"[8]).

Another source of funding came from the Central Intelligence Agency (CIA). Distinct from much of the CIA's investment in technology at the time (its research and development initiatives were focused on building reconnaissance planes and spy satellites), as Sutherland clarifies in

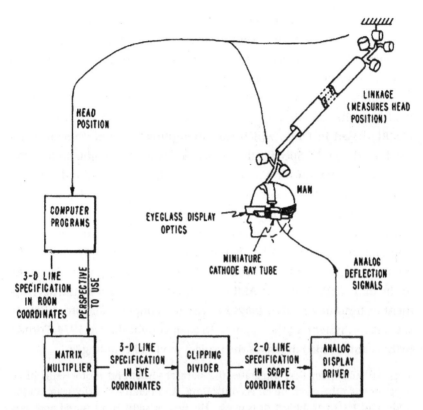

Figure 5.2 A diagram of Sutherland's ultimate display from 1968.

a 1996 lecture,[9] the CIA's funding had a different purpose. The CIA had been allocating money to aboveboard research projects to offset negative perceptions to do with its maleficence in operating domestically in the 1960s (specifically, infiltrating left-wing student activist groups). In Sutherland's telling of events, being bankrolled by the CIA was relatively benign (although the backlash from student activists at Harvard showed that not everyone agreed).

For Sutherland, the goal of the head-mounted display his team developed was to take Bell's experiment with augmenting helicopter pilot vision and substitute the camera for a computer, which would enable the wearer of a head-mounted display the ability to "view a mathematical world of our own choosing. We could see anything we wanted from any angle. And that would make it easy to understand complicated shapes."[10]

It is in this sense that Sutherland's ultimate display embodied many of the tenets common to mid-century defense programs like SAGE, where watching the skies for aerial threat through computerized systems would come to augment fallible human perception.

Running parallel to the work done at Harvard and with Bell by Sutherland was the US Air Force's Aerospace Medical Research Laboratory (AMRL) based in Ohio, which was attempting to refine "aircraft-man-machine design."[11] Specifically, the lab was researching how to improve visual augmentation for aircraft pilots—for both training and eventual incorporation into aircraft design. Unlike Sutherland—who was principally interested in a problem of human-computer interaction—the AMRL's chief goal was more applied in nature: seeking to enhance the optical acuity of pilots. A key figure here was Thomas A. Furness III, an electrical engineer in the Air Force—who began working with the AMRL in the mid-1960s. With the AMRL, Furness was involved in the development of head-mounted displays to "visually couple"[12] air force pilots to the weapon systems of their planes. In a paper published in 1974, Furness (with coauthor Birt) describes the project of visual coupling as

a special subsystem that integrates the natural visual and motor skills of an operator with the machine he is controlling. An operator visually searches for, finds, and tracks an object of interest. His line of sight is measured and used to aim sensors and/or weapons toward the object. Information related to his visual/motor task from sensors, weapons, or central data sources is fed back directly to his vision by special displays so as to enhance his task performance. In other words, he looks at the target, and the sensors/weapons automatically point at the target. Simultaneously with the display, he verifies where sensors/weapons are looking. He visually fine-tunes their aim, and he shoots at what he sees.[13]

Furness would subsequently be involved in developing—throughout the late 1970s and early 1980s—the Visually Coupled Airborne Systems Simulator (VCASS). Originally intended as a flight simulator, the headset became a way to communicate information to pilots—a "fully immersive three-dimensional circumambience of graphical information superimposed over the real world," as Furness describes it.[14] VCASS was completed in September of 1982. The centerpiece of the system was a helmet consisting of two CRT displays—"one for each eye—hanging from a platform and connected to eight mainframe computers running computer graphics software."[15] VCASS filled several rooms and used so much electricity that Furness joked he "had to tell Dayton Power and Light" whenever he was going to power up the system.[16]

Another key figure within the postwar era of military technoscience was Raymond Goertz, an engineer for the Atomic Energy Commission at Argonne National Laboratory in Illinois (a laboratory that emerged from the University of Chicago's involvement in the Manhattan Project).[17] Goertz's work was particularly influential in the development of *haptic interfaces*—relating to the way that the hands could be technologically mediated to manipulate objects, both real and referents in digital space. Working in the lab's remote-control engineering division, from 1948 to 1949, Goertz developed a system referred to as the *master-slave manipulator*—a set of mechanical arms that transmitted feedback to the hands of a remote operator (initially, via a system of steel pulleys and cables, then from 1954 on, through a system translating mechanical forces into electrical signals that could be communicated via alterations in current to the motors in the arms).

The intended outcome of Goertz's device was to advance techniques for the manipulation of radioactive material, seen as a necessity by the United States in the development of nuclear armaments in the context of heightened postwar tensions with the Soviet Union. Because of this, the device needed a high degree of precision. In response, Goertz developed and applied the principle of *degrees of freedom*. The motion of the slave arm, to accurately simulate the full range of hand gestures, must possess six independent degrees of freedom: three of translation (movement in 3D space in the X, Y, and Z axes—facilitating the movement of the hand up and down, left and right, and back and forward) and axial rotation (movement across the U, V, and W axes—enabling the hand to position and exercise grip). Degrees of freedom is a now-standard measure of how many axes of movement a gestural tracking device can utilize (and most devices today provide six degrees of freedom).

For Sutherland, Furness, and Goertz, the goal of their research and development was to leverage the visual and haptic affordances of their respective interfaces to mediate the modes of experience and perception in the manners desired by state defense departments, defense research labs, and military contractors. These military entanglements with foundational VR technologies set the stage for the fantasies of simulation and violence that underpin its use by the state today.

REAL WARS ON VIRTUAL BATTLEFIELDS

The simulated modeling of wartime operations where VR emerged remains a prominent use case now that VR systems do not require rooms of computing equipment. Where the ultimate display was used to enhance the optical acuity of helicopter pilots, and the master-slave manipulator to simulate the gestures of a human hand to manipulate dangerous radioactive matter, VR's multisensory affordances (specifically those of mediating visual and haptic senses) have been put to work as a delivery mechanism for military training exercises, simulating hypothetical—and imagined realistic—scenarios of warfare.

The US Army and Department of Defense, for instance, increasingly see VR as a major component of what they term the synthetic training environment (STE), a yet-to-be-realised "collective, multi-echelon training and mission rehearsal capability for the operational, institutional and self-development training domains."[18] In a press release, the US Army frames the need for the STE with reference to the 2003 ambush of a US convoy in Iraq (after it had taken a wrong turn into hostile territory based on receiving incorrect directions), resulting in the deaths of eleven soldiers. As the press release notes:

Factors leading to this wrong turn included a lack of equipment, a lack of maps, poor judgment, and a lack of training . . . Field Manual 7–0, Train to Win in a Complex World, highlights that planning and rehearsing lead to better execution. Simulations enable Soldiers to plan and rehearse events prior to executing operations in a live environment. The 507th Maintenance Company may have avoided this tragedy if it had virtually rehearsed this convoy route before executing the mission.

Imagine an environment in which sustainment Soldiers can put on a pair of virtual or mixed reality goggles and find themselves in any country in the world and on the same type of terrain they will operate on in the near future. In this environment, they are connected with their supported maneuver force and joint and coalition partners, and they have the ability to rehearse the sustainment plan developed for the mission numerous times before they execute it.[19]

Grounded in a fantasy of "fixing" fallible humans—of VR as a posthuman extension of human action and cognition—the STE is imagined by the Department of Defense as *scalable* (can be implemented across a large userbase), *mutable* (able to be updated with current training regimens, able to meet changes in military hardware), and *complete* (training simulations that capture "the data that we're leaving on the floor,"[20] incorporating it into training management and assessment). The fantasy of the STE is not of the AI swarm of autonomous robots that removes human agency,[21] but of the centralized management of multidomain operations—allowing military personnel to maintain readiness and increase lethality.[22] The STE, in other words, represents a fantasy of creating more efficient killers.

While much of the STE is a hypothetical vision of the future of command and control, there are aspects that do concretely exist. In bids for lucrative contracts to supply hardware and software for the STE, defense firms have developed prototype simulations—putting forward their own fantasies of what VR can do for military training. Raytheon, for example, has developed a prototype VR-based simulation of its FIM-92 Stinger missile.[23] Elsewhere, under a funding scheme to develop medical simulators for the STE, the defense department contracted the firm HaptX, a company specializing in "goggles and gloves" simulations. HaptX pairs an HTC VIVE VR headset with the company's own proprietary gloves. As the company puts it, the simulation creates an illusion of reality through a combination of the headset's visual and auditory affordances and through tactile actuators in the gloves (which create physical sensations throughout the hand), primed to the visual and auditory outputs of the VR simulation—thus creating the sensation of feeling in the hand. The result, HaptX claims, is a "hyper-realistic . . . and fully sensored simulated weapons for the synthetic training environment."

VR elsewhere operates in the simulation of military hardware. For the American firm Kratos—which is funded by a $17.6 million contract from the US Air Force—VR is a central component of its simulators for vehicles and armaments. Kratos's aircrew training simulator, for example, provides haptic simulation of aircraft armaments, paired with the auditory and visual simulation of a VR headset.[24] Other military across the world have begun to deploy VR to emulate the use of military equipment. The Australian Army, for example, uses the Protected Mobility Tactical Trainer—a system that combines emulated vehicles and weaponry with VR. In one demonstration, we see a vehicle with a mounted gun operated by a trainee adorned with a VR headset. Further to its simulational fidelity—in a way that chimes with the sales pitch of many other technological devices of the moment as "smart" and networked—the tactical trainer is described in an Australian Army press release as "interoperable" with the wider digital networks comprising the modern war machine.[25]

Beyond training—particularly in the use of weapons and combat—VR simulations operate as a mechanism to empower the broader economy of war, particularly in the process of procurement. VR-based simulations are extensively used at venues such as defense trade shows. Trade shows are an important venue for defense companies, particularly in that they provide a way for vendors to "sell" state defense departments on technological innovations, attracting often lucrative contracts. VR here is not the saleable product, but rather the tool through which defense companies can simulate their products—which would otherwise be unfeasible or unsafe within the trade show environment. One example we observed, at the 2017 Defense and Security Equipment International trade exhibition, was defense and security company Saab partnered with the company HapTech (not to be confused with the aforementioned HaptX), which specializes in creating emulations of military equipment via a combination of a HTC VIVE Pro, an OptiTrack motion capture system, and a digital "prop" resembling some kind of armament.[26] In this case, HapTech's simulation was used to demonstrate a Carl Gustaf M4 antitank gun to the trade show's attendees. Simulation—and VR—are not just a product of the military-industrial complex but are key in fueling it.

Adjacent to VR is the wider industry of mixed reality computing interfaces being developed for the military—interfaces that incorporate a broader range of physical elements into digital space. Perhaps most notable in this space is Microsoft, which in 2019 secured a $480 million contract with the US military to develop a mixed reality system in the style of its commercial HoloLens headset for use for military combat and training (receiving a subsequent $22 billion to further develop and supply these headsets over the next two years).[27] Microsoft's patent[28] for a military-grade mixed reality headset imagines how wearable mixed reality technologies will enable real-time capture and relay of information among soldiers, but also between soldiers and reconnaissance technologies such as drones—rendered within the headset's interface, enabling more efficient targeting, tracking, and killing (or, increasing lethality). We are presented with an image that encapsulates the United States' imperialist ambitions of conquest in the Middle East and how they are coupled with the fantasies of the technological extension of human perception. We see an Arab man being tracked and read as a threat through the HoloLens's iris recognition—rendered as such through information stored in databases, transferred to the headset at light speed—a system that feeds into and shapes the anticipatory capacity of the device's user in real time, as shown in figure 5.3.

The fantasy here is one of the systematic coordination of soldiers, computer-based systems, and the sensing capacities of the device. How soldiers engage in conflict unfolds through mixed reality interfaces that capture information about external environments and the organic and inorganic things in them. These are framed as minimizing the need for soldiers to make split-second decisions in combat. Decisions for using lethal force are distributed between the soldier and the data centers feeding information into the mixed reality device's heads-up display.

FANTASIES OF COMMAND, COMMUNICATION, AND CONTROL

Beyond the military's fantasy that VR's haptic and visual affordances can be mobilized for more effective simulations and training—a second military fantasy is of VR as a mechanism for modeling or visualizing information. Indeed, computers have long used graphics to index vast volumes of data to make it perceptible by humans (rather than simply something to be parsed by machines)—notably, by members of the military. As media historian Bernard Dionysius Geoghegan puts it, computer graphics (and the notion of the visual in computing more generally) "gave rise to the conception of computing as a multimedia collaboration among humans and machines"[29]—originally, a means to convene modes of attention necessary for aerial defense in the Second World War. For Geoghegan, "The screen permitted a redistribution of human perceptions in ratios strategically matched to the challenge of jet propulsion and modern munitions"[30]—a way to bring humans in line with the space-times of war. With Cold War–era computing systems like SAGE in the United States, computers were a mechanism for facilitating the task of creating a

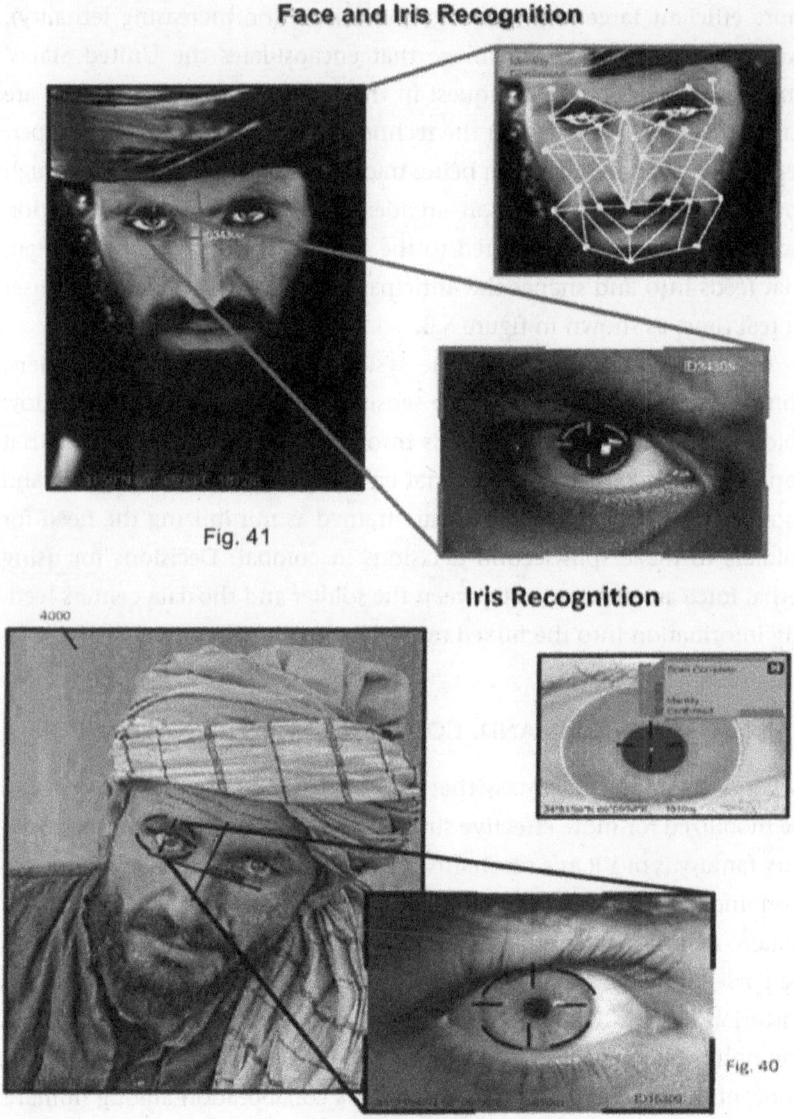

Face and Iris Recognition

Fig. 41

Iris Recognition

Fig. 40

Figure 5.3 Facial and iris recognition capabilities for military-grade augmented reality (AR), as imagined in Microsoft's patent.

nationwide command, control, and communications system (connecting radar facilities, Air Force bases, and other operational units).

Different from two-dimensional displays of information, VR is seen by defense departments and contractors alike as a new computing paradigm, one that can convene the modes of attention necessary in contemporary battlespace, across haptic and visual registers. For instance, VR forms a crucial layer for simulating and visualizing data in the US Air Force's Advanced Battle Management System (ABMS).

The Air Force describes the ABMS as an "internet of things" that would use computation and automation to mediate decision-making, falling under the Department of Defense's Joint-All Domain Command and Control (JADC2) initiative—a $950 million program drawing on data from across the branches of the military to create a centralized network, sharing intelligence, surveillance, and reconnaissance information.[31] The ABMS is made up of an ensemble of networked and sensing technologies, and VR is one part of this, provided by companies such as Immersive Wisdom. Immersive Wisdom is a start-up founded in 2016 that specializes in "3D remote virtual ops"—essentially, data visualization and simulation—that "enables real time geospatial collaboration and intelligence across both disparate users and data sources."[32] Initially, the company attracted funding from the CIA's technology venture capital firm, In-Q-Tel (which makes investments in firms that provide solutions to the CIA's unclassified objectives).[33] More recently, in 2020, it received $190 million from JADC2.[34] For Immersive Wisdom, the multisensory affordances of VR make it richly resourced for visualizing and manipulating data. As the company puts it: "Immersive Wisdom offers a 3D remote virtual ops center platform for distributed and disaggregated operations that allows geographically dispersed personnel to effectively collaborate and act without having to be physically together . . . sensor feeds, enterprise applications, maps, 3D data, geospatial sources, and video streams into a synchronized real-time, interactive virtual 3D operations center."[35] VR is presented as a solution to a problem of incompatibility between human and machine—where data operates at speeds and times, and at volumes, that are beyond the capacity of the human user to process. VR simulations provide users with the opportunity to observe, capture, and transmit data to other interfaces and personnel, collaboratively identifying and responding to potential threats, thus extending human capacity to more effectively (read, lethally) act.

Another firm in this space is Anduril, a defense tech contractor for the US government specializing in the development of AI systems incorporated into VR, drones and large sensor towers. For the most part, the "threats" these systems purport to track refer to asylum seekers from Central and South America attempting to cross the border between the United States and Mexico, but they are also used to defend military bases in the United States. Indeed, it is here that VR's utility dovetails with the broader imaginary of the drone—one that seeks to extend the sovereign power of the West through a doctrine of "preempting" threat through mediated vision.[36]

Anduril's relation to VR is perhaps best known through the company's founder, Palmer Luckey—who, as you will recall from chapter 2, was also the founder of Oculus. Luckey founded Anduril after being fired from Facebook in 2017.[37] In the years prior to Oculus and Facebook, Luckey worked as a technician at the Mixed Reality Lab of the Institute for Creative Technologies, a Department of Defense–funded research center located at the University of Southern California. In 2014, the lab developed an information dashboard for the US Navy—an initiative called BlueShark—that visualized information using Luckey's Rift prototype. As Mark Bolas, director of the Mixed Reality Lab, notes, these interfaces were seen as a stopgap by the navy for AR and holographic displays—which didn't (and still don't currently) exist at scale, providing a site for small-scale experimentation with what these kinds of interfaces might look and feel like.[38]

Some of Anduril's use cases represent an updating of these experimental visions of virtual command and control. In a 2020 feature for WIRED magazine, Tom Simonite describes the process of using Anduril's AI tracking system, Lattice, with a VR headset to visualize imminent threat to a military base in New Mexico: "A member of the US Air Force donned a virtual reality headset and scanned a 3D map of a desert landscape. He saw a speeding object that algorithms warned was likely a cruise missile. The airman considered the data, then used a hand controller to send out an order. . . . When the mock missiles started flying, Anduril algorithms tracked the foreign objects and alerted Lattice users that the system had detected what appeared to be a missile."[39] Simonite goes on to describe the rendition of Lattice: "The software can be used on conventional displays, but the main operator during the exercise used a VR headset from Facebook's Oculus division, a descendant of the technology Anduril cofounder Luckey sold the social network. Inside the headset, Lattice displayed a 3D map of White Sands Missile Range with aircraft and other objects highlighted."[40]

Beyond the defense of US military bases, a major market for Anduril is border control (its technology has been contracted by US Customs and Border Protection). Gazing over borders in Arizona, California, New Mexico, and Texas, a VR headset provides the user a view of the border and of targets being tracked and identified by the Lattice AI system. Lattice forms a virtual wall, a lower-cost and more logistically sound solution than Donald Trump's call to build a "big, beautiful wall" between the United States and Mexico. Through Anduril's Lattice system, a combination of autonomous AI and VR-headset-wearing human operators enforce restrictions on freedom of movement, restrictions that remain in place today under the Biden administration's continued efforts to deter, arrest, and incarcerate some of the world's most vulnerable people. (This has been redoubled through Biden's US Citizenship Act, a bill that emphasized the use of "smart" technology for border security. VR is part of the suite of technologies that supposedly enable border forces to "responsibly manage"[41] the border.)

In this way, we see how VR is part of a neoliberal technosolutionist agenda for the "problem" of migration. Technologies in Anduril's stack, such as VR, that facilitate the sorting, tracking, and classification of migrant activity at "machine speed" and with "unparalleled confidence . . . by turning data into information, information into decisions, and decisions into actions across tactical and strategic operations"[42] are pitched as humane "solutions," yet have only exacerbated human suffering—leading to injury and, in some cases, death.[43]

Technologies like those of Anduril and Immersive Wisdom enact a technophilic fantasy of omnipotence—of being an all-seeing eye, gazing over large volumes of information derived from vast networks of sensors. For these fantasies of VR as a mechanism for command and control, information—fed through the headset—shapes the capacity for thinking, feeling, and acting in the future. This imagined "distributed" nature of military warfare is framed as enhancing accuracy and accountability. Indeed, this promise has clear parallels to the wider history of militarized technics and sensing propagated by the West (and perhaps most clearly exemplified in the drone): one based on mediated techniques and "anticipatory logics of developing a pre-emptive mastery of the territory and its potential threats,"[44] sensing and modeling—through technology—perceived enemy potentiality. What is being advocated—much as with the military-industrial framing of the drone—is the strategic necessity of adopting the deployment of data-driven systems as a military doctrine of preemption and automated classification of threats.[45]

FANTASIES OF CYBERNETIC COPS

Political and cultural theorist Paul Virilio famously wrote that we live in a state of permanent (or "pure") war.[46] By this, he meant that there is an increasing "perversion" of any clear-cut distinction between civilian and military institutions, and by extension civilian and military life. War spills from once-distant geographic sites of military conflict to the everyday. The point for Virilio is not simply to suggest that consumer technologies filter down from military technoscientific experimentation, but rather that following the Second World War, economy and society in the West were permanently reorganized around the aims and logics of advancing military power. The capacity to rapidly mobilize for war requires the gears of the wartime economy to never stop turning. War subsequently shifted from the strategies and tactics of battlespace to a question of logistics, a broader strategic directive involving the total reorganization of industrial society, such that it could rapidly produce rockets, missiles, and other armaments (or, more recently, computational, algorithmic, and sensing technologies).

It is in the context of *pure war* that the technologies, techniques, and logics of militarization spill into the everyday. One setting in which this happens is through the increasing militarization of technologies used in the governance of civil societies, such as by police forces, augmenting the domestic disciplinary arm of the state. Through the diffusion of military logics and technologies into everyday life, war and militarization are made "banal,"[47] as political economists Greig de Peuter and Nick Dyer-Witherford put it. Writing on this topic, geographer Stephen Graham refers to this as the "militarization of everyday life": the "insidious"[48] creep of both technologies that track, sort, and profile, and logics that have increasingly come to characterize approaches to urban governance. Cops see urban environments as sites of "threat"—as they are outfitted with rugged tactical armor, high-powered assault weapons, and facial recognition software. Logics of militarism—flowing through technology—extend beyond the spatiotemporal bounds of war and into quotidian life.

For AR firms that provide software and hardware to police, an enduring fantasy is of creating "smarter" cops who operationalize real-time data flows in ways that mirror common depictions of cyborgs in popular works of fiction. For example, Vuzix—a major manufacturer of augmented reality headsets, which has developed applications that incorporate facial recognition (working with companies like the controversial Clearview AI)—makes the (curious) comparison between a wearable augmented reality headset and Paul Verhoeven's 1987 film *RoboCop*. As the company states on its website: "While the dystopian society envisaged in *RoboCop* is nothing to emulate, the innovative tools its main character uses to protect those in need is within reach."

Like Palmer Luckey or Michael Abrash in their desire to recreate *The Matrix*, Vuzix seemingly misses Verhoeven's subtext in *RoboCop* of social satire and commentary on the corporatization of urban governance in light of growing privatization of social services under Reagan-era austerity. It also seems lost on Vuzix that, as a purveyor of technologies of violence and control to police forces, it plays a similar role to Omni Consumer Products—the nefarious corporation and antagonist in the film, a company that facilitates and seeks to profit from the chaos of a dystopian Detroit through the provision of extremely harmful technologies to police forces. As we have pointed out elsewhere in a more comprehensive review of AR police tech,[49] the goal of many of these AR applications for policing is to mediate how police think, feel, and act while *doing* police work. The point is that they provide police with a kind of anticipatory edge over the "threats" of the urban environment, feeding and visualizing streams of data—such as those derived from facial recognition systems.

VR policing applications tend to imagine a different kind of cybernetic cop. In contrast to AR, the simulated modeling of the world through VR seeks to shape the future actions of police differently; a VR simulation presents the opportunity to shape how police think, feel, and act *before* they do police work. There is a different kind of anticipatory logic at play. It is one of instilling the user with preparedness—one rooted in the assumption (a largely fallacious one, as outlined in chapter 3) that VR has the capacity to powerfully shape how people think, feel, and act.

One notable use case has been the use of VR as a simulation tool for training police forces, currently being rolled out across the world—such as in the United States, the United Kingdom, and Australia.[50] One market-leading firm in this space is Axon (previously Taser), a company that chiefly develops technology and ostensibly nonlethal weapons for military, police, and civilians (most famously the Taser electroshock weapon, widely used by police in the United States). In 2017, following the rebranding of Taser to Axon, the company began developing bodyc-ams for police forces, sold on the premise of transparency, accountability, and more ethical policing—a response to a spate of high-profile police killings of predominantly Black and brown people in the United States driving the adoption of these bodycams. Axon's cameras were a means to "capture truth"[51] (while also operating as a "rugged communications bea-con"[52]), allowing the communication of information between cops—a tool for further weaponizing police under the guise of security.

Extending this reformist vision of policing, in 2021 Axon began devel-oping VR simulation tools (as part of a VR-based learning management system) to be sold to police departments—framed as providing both weapons skills and empathy training. As Axon put it: "Combining the HTC VIVE Focus 3 VR headset and the VIVE Wrist Tracker from indus-try leading partner, HTC VIVE, Axon VR provides an all-in-one, portable platform that's simple to deploy and use, with no extraneous hardware, time or space constraints. The VR training provides connectivity for both in-person and remote experiences, creating an on-demand platform that can be accessed anytime."[53]

Axon also offers simulator training for weapons, including its own Taser energy weapons, as well as firearms training—equipping users with an accurately modeled and weighted handgun, outfitted with sensors such that its movements are modeled in the VR simulation. The risks of making cops *more* lethal in the United States—a country with a particu-larly murderous police force—should be obvious. But Axon's selling point for its VR is that "unlike traditional simulators that only offer use-of-force training, Axon VR enhances an officer's ability to de-escalate many of the most common calls for service." Put differently elsewhere, Axon offers "empathy training"—providing officers with "immersive content designed to encourage critical thinking and de-escalation." In their view, de-escalation training aims to bring about "improved civilian interactions" and to help "rebuild the fractured relationship between cops and communities." The company sees itself as "spearheading the dialogue that helps heal society."[54]

A particular point of focus for Axon in its promotional materials is a VR-based module simulating an encounter between a police officer (played by the user) and a person (modeled in the system) with various forms of cognitive impairment (with modules including autism and schizophrenia—options, as one journalist notes, that are "laid out on a menu screen like the levels of an early 2000s platformer game"[55]). Axon reports that its VR training modules have been adopted by over one thousand police departments across North America. For Axon, the modules are "not about stopping criminals"; rather, they are about equipping first responders with the skills to "recognize and successfully deal with a range of mental and psychological conditions." Further, "the goal of these modules isn't about 'catching the bad guy' but rather work to ensure the safety of the subject as well as the officer."

In a similar fashion, Google's Jigsaw division—which focuses on "threats to open societies"[56]—has developed Trainer, a VR platform for "adaptive scenario-based training."[57] As Google describes it: "Recent advances in virtual reality (VR) have demonstrated the potential for technology to create scalable opportunities for law enforcement and other public safety professionals, particularly in the realm of education. In training contexts, VR can create a uniquely immersive experience, employing heightened tensions to build critical skills in an environment that mimics the same physiological responses as those generated in real-world interactions."[58] VR here is part of an evaluation assemblage that hinges on Google's natural language processing AI, Dialogflow. In this way, Google claims that its software can categorize likely user intent based on speech inputs to the system's interactive voice-response system (where users communicate with virtual characters).

For both Google and Axon, contra a wider range of mixed reality policing technologies and other policing tech (e.g., "predictive policing"), training is framed as *explanatory* rather than *predictive*. The outcomes of how police perform at training simulations have—such as in the case of the Sacramento Police Department—formed part of evaluative criteria in training, a way to divine which cops are "good" and which are "bad" (and a data point for backing up arguments about police reform). It is hard to take seriously such propositions of VR as a techno fix for deeply sedimented, structural issues of racism, ableism, and authoritarianism that pervade police forces and underlie the violence enacted by the particularly murderous police forces in the United States (which currently form the largest market for these training simulations). Additional training—presented as a common suggestion by those arguing for police reform (as opposed to abolition)—is unlikely to dislodge the entrenched "warrior" culture in police departments across the United States, characterized by a belief in the "thin blue line," the worldview commonly held by cops that they're the only thing keeping society in check and must do so at any cost. In contrast to the widespread dissent by activists and critical abolitionist scholars, who question the moral legitimacy of policing and the carceral state, policing technologies like those of Axon and Jigsaw are cynically trying to profit from police forces that need to rebrand in the face of a crisis of legitimacy.

Writing in the context of AI, Dan McQuillan notes that "from a state point of view, the arguments for adopting AI's alleged efficiencies at scale become particularly compelling under conditions of austerity."[59] Indeed, this is very much true of VR's solutionist application in policing contexts. Axon's de-escalation solutions —what it describes as "scalable and affordable"[60]—could also be seen as an indictment on the failures of the intensification of the neoliberal "small government" and the growing austerity measures resulting in the dismantling of public services, the slashing of social welfare, and an uptick in spending on military, police, and prisons. Axon reinforces a prerogative to defund social services and replace them with community policing initiatives. This further reinforces the fact that police are now regularly called on to de-escalate behavioral health crises and distress. As Leanne Dowse and colleagues put it, "Police officers have become the 'carers' of last resort, and the leading agency in 'managing' disadvantaged people with disability."[61] Relatedly, and in light of claims that much of policing involves the control and regulation of "difference" via the criminalization of relatively nonserious behaviors and activities (particularly by marginalized communities), VR-based training models—which rely on a necessary reduction of complexity (e.g., glossing over how cognitive disorders often co-occur with substance abuse issues or with other cognitive disorders)—have the potential to exacerbate harm.

Such VR systems of simulation—which effectively work to train police to target, profile, and criminalize difference (along the lines of race or disability)—feed into a wider societal problem, what abolitionist scholar Jackie Wang calls *carceral capitalism*.[62] As Wang argues, with the technology-driven nature of hypermilitarized and surveillant policing—where new technologies facilitate the criminalization and prosecution of relatively nonserious offences—the dynamic has proven incredibly lucrative for the private sector firms in the carceral industry (such as those that provide the telecommunications systems for prisons, for which they charge exorbitant usage fees). The carceral industry sees new opportunities for profit in VR. One particularly egregious example is Global Tel Link (GTL)—a prison contractor that provides telecommunications systems and payment services to prisons in the United States. In 2017, GTL filed a patent for a "system and method for personalized virtual reality experience in a controlled environment."[63] Put plainly, GTL wants to charge prisoners to use its VR software, allowing the incarcerated to, "for a brief time, imagine himself outside or away from the controlled environment."[64] VR is at once the mechanism for training cops to target and persecute difference and the mechanism from which a parasitic carceral technology industry seeks to derive profit from that same oppression.

Fantasies of simulation—of VR's capacity to model real-world phenomena—are often considered in neutral terms. Yet simulation is neither made nor deployed in a vacuum. It is closely entangled with the values, aims, and logics of the institutional systems within which it is embedded. For all its supposed novelty, as this chapter shows, VR entrenches forms of thought and action that reinforce the status quo. This is particularly apparent when adopted by police forces and militaries—where a fantasy of simulation might be more accurately construed as a fantasy of violence: a technology bringing about injury, harm, and even death, or exacerbating existing forms of inequality, discrimination, and bias.

6

FANTASIES OF PERFECT DATA

At its most basic, a VR headset is a computer display with a head-position sensor so that the display can change based on movements of the head. This was the capability that Ivan Sutherland first demonstrated in 1968, with his Sword of Damocles tracking system, named as such because the intimidating mechanical system was suspended above the user (see chapter 5, figures 5.1 and 5.2).[1] Sutherland also described in his 1968 paper an "ultrasonic head position sensor," an approach that instead used ultrasonic transmitters. These transmitters broadcasted at different frequencies, and receivers in a square array mounted on the ceiling tracked their position. However, because the wavelength of the transmitters was around a third of an inch, this method of tracking was never accurate enough to function adequately. The experience of Sutherland's display was more similar to augmented reality devices like Microsoft's Holo-Lens because the display was actually see-through: no screen at the time offered sufficient resolution to be held close to the eyes. With the basic principles in place, but none of the sufficient technology to use it in any practical applications, the fantasy of VR had to wait for the technology to catch up.

For a little while in the 1980s, it seemed as if the technology had caught up to the fantasy. The US Air Force (grappling with the challenge of training pilots to fly increasingly complicated machines) and NASA (which had the challenge of putting humans in even more dangerous places) drove research into VR. Michael McGreevy—a researcher at NASA's Ames Research Center—launched a new VR research project in 1985, after noticing that the Citizen watch company's LCD displays in mini consumer televisions were small enough to fit into a head-mounted display, and with high enough resolution to be positioned three inches from the eye.[2] NASA's fantasy was of teleoperated space exploration, with human controllers using VR to control robots across the solar system.[3] Only a year later, McGreevy launched the Virtual Visual Environment Display (VIVED) headset at the 1986 Consumer Electronics Show (see figure 6.1). Each system, costing thousands of dollars, used electromagnetic sensors to determine the changing positions of the headset inside an electromagnetic field. The sensors could accurately detect their location within the field, but it was subject to magnetic perturbations in the environment,[4] which is not so great for use in space.

Under Scott Fisher, NASA also developed the Virtual Interface Environment Workstation (VIEW; see figure 6.2)—a similar headset, but with a focus on going beyond the VIVED's focus on display. This incorporated several other advances that have become fundamental to making VR feel "real." The headset had 3D binaural sound, meaning that the audio through the device's stereo headphones changed depending on the orientation of the head, and it incorporated a gesture-based natural user interface using a modified version of the DataGlove that was developed at VPL, a VR research and development company founded by Jaron Lanier in 1984. The DataGlove used a patented optical flex sensor to track if fingers were bent or straight, and more sensors ensured that the absolute location of the hand could be tracked and therefore displayed to the user in VR.[5] An added benefit of this is that if a VR system knows the location and orientation of both your head and neck, VR can approximate what the rest of your body is doing, and represent that in the virtual world too. Each finger of the DataGlove had tiny vibrators to provide haptic feedback, which could provide a sensation of touch. One of these VR systems—and the supercomputer needed to power it—cost tens, if not hundreds of thousands of dollars, but for NASA the added presence and immersion of these additional interfaces not only promised to fulfill a fantasy about training and simulation, but also could enable telepresence and remote control that would make space exploration cheaper and safer— cheaper than a spaceflight, at least.[6]

The consumer hype around VR reached its first peak with the announcement of the Sega VR at the 1993 Consumer Electronics Show, the doomed VR gaming peripheral we discussed in depth in chapter 2. One of the

NASA TechBriefs

Transferring Technology to
American Industry and Government

July/August 1988
Volume 12 Number 7

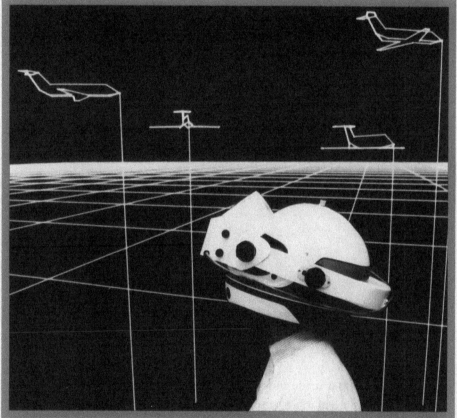

NASA's Virtual Workstation Shapes A VIVED Reality

Figure 6.1 NASA's VIVED headset, first demonstrated at the 1986 Consumer Electronics Show. The VIVED system focused on the potential benefits of immersive and 3D displays and on more immediate possible implementation.

Figure 6.2 NASA Scientist Sally Rosenthal demonstrating the VIEW VR System in 1989, including the DataGlove interface developed by VPL. *Source:* Photograph by Wade Sisler.

developments that made the Sega VR possible as a sub-$500 device was its "sourceless orientation sensor," invented by Mark Pesce. Rather than detecting changes in orientation in a human-generated electromagnetic field (as NASA's VIVED headset did), Pesce's invention elegantly detects changes in orientation and elevation against earth's electromagnetic field.[7] Movements of the head could now be tracked for less than one dollar,[8] but computer graphics and screen technology were still decades away from being able to realistically fool the body's other senses.

Palmer Luckey's Oculus Rift system—the one first demonstrated at the 2012 E3 gaming convention, which would catalyze much of today's interest in VR—operated on essentially the same principles as the Sega

VR. What had changed to make the Rift possible was the size and speed of technology, principally thanks to advances made in smartphone manufacturing; in fact, the Oculus Rift Development Kit 2 used a modified version of the Samsung Galaxy Note 3 screen. In an interview with the Verge website in 2014, Luckey notes: "Here's a secret: the thing stopping people from making good VR and solving these problems was not technical. Someone could have built the Rift in mid-to-late 2007 for a few thousand dollars, and they could have built it in mid-2008 for about $500. It's just nobody was paying attention to that."[9] The consumer version of the Rift released four years later used external sensors—called the Constellation—which were in fact just webcams placed in front of the user to track a "constellation" of infrared LEDs on the headset. These LEDs blinked in specific patterns, allowing the software to deduce the position of the headset using the same fundamental method as Sutherland's ultrasonic head-position sensor, but with the necessary accuracy. Combined with inertial measurement units (IMUs) inside the headset, this gives VR software sufficiently high-frequency updates about where the display is, in real time. Rather than using gloves, the Rift used controllers inspired by video game consoles, whose vibrations provided a sufficient haptic sensation of touch, and had built-in microphones and stereo speakers to provide binaural sound. Together, these advances provided the seamless and convincingly real experience of a virtual world that has reinvigorated excitement for the technology.

But this type of VR remained tethered to a single location. As we discussed in chapter 4, for Meta, this restriction was seen as a barrier to VR's widespread adoption and their capacity to build an ecosystem over which they could have control. As early as 2016, Facebook was working on an all-in-one headset,[10] which ultimately became the Oculus Quest (released in May 2019). Unlike the Rift, the Quest was not tethered to a PC that handled the processing, and it did not require the placement of sensors around one's room to track movement. The Quest functions via *visual-inertial simultaneous localization and mapping* (VI-SLAM), a computational method of constructing a digital map of the environment that a device is located within. The Quest is reliant on a combination of IMUs and outward-facing cameras that feed into a system that Oculus calls Insight. Image data from the headset helps generate a 3D map of the room that is updated in real time. In combination with the IMUs that retrieve "linear acceleration and rotational velocity data" from the headset and controller, the system can establish where objects are relative to other objects in space. As Meta's engineers outline in a post on the company's AI blog,[11] the goal is for the system to know where the device is, relative to where it was, and also where it is moving to. The speed of this process supposedly alleviates the visual jitter and discomfort that typically emerge in VR when there is a delay between user input and the generation of the VR image.

Reflecting the quite profound impact that deep learning and AI are having in the field of computer vision, Meta engineers updated the Quest in 2019 with an impressive hand-tracking system that uses just the black-and-white images from the camera to create a 3D model of the hand, with fingertips and joints: no DataGlove required.[12] This allows the Quest to bring your hands into the virtual environment without the use of instrumented gloves or controllers, creating an even more seamless and immersive experience.

The way VR's sensemaking has been externalized via systems like the Oculus Quest is also part of a trajectory of software development that might later enable ubiquitous AR devices, something that Mark Zuckerberg has spoken candidly about in media interviews and earnings statements and is reflected in recent EULA changes. The Quest's externally facing cameras are used in its "passthrough" system—which allows the user to swap into a grayscale view of the physical environment around them. This hints at future AR-like experiences with the all-in-one Quest, but right now it is only used to safely demarcate the boundaries of the play space so that, for example, users don't punch a wall while playing *Super Hot VR*.

These data-collection capabilities combined—closely tracking our head movements, hand movements, and what the physical world around us looks and sounds like—mean that modern VR devices are collecting more data about us, and the physical world, than ever before. Jeremy Bailenson, founder of the Stanford Virtual Reality Lab and cofounder of VR company Strivr, lucidly describes the modern-day data-collection capacities of VR: "In 2018, commercial systems typically track body movements 90 times per second to display the scene appropriately, and high-end systems record 18 types of movements across the head and hands. Consequently, spending 20 minutes in a VR simulation leaves just under 2 million unique recordings of body language."[13]

In a 2021 podcast interview, Zuckerberg asks, "How do you pack even more sensors, to create a better social experience, into the device?"—hinting at Meta's exploration of other sensing capacities for VR, from heart rate sensors to brain-computer interfaces. What are the implications of this expansive data collection?

USING AND ABUSING VR DATA

Many VR companies are already deploying or imagining opportunities for what might be possible from the analysis of this data. VR fitness games will be able to draw on heart rate data and changes in our movements to adapt instantly to our level of fitness, ensuring that the physical challenge keeps us at an optimal level of exertion for the most impactful results. Classroom learning applications could personalize a student's learning experience by recognizing moments of confusion—identified in eye-tracking data—and using that knowledge to reiterate a lesson, or to try explaining the information in a slightly different way.

The more computers know about us, the more assistive agents like these can support us in the tasks we're working on. Researchers at the University of Melbourne, for instance, used eye-tracking data from users playing the board game Ticket to Ride to create an assistive agent that could help players make gameplay decisions, helping improve their decision-making by highlighting things they hadn't noticed yet.[14] This works because where we look and how we look at things offers insight into what we're thinking.

Strivr is at the forefront of these kinds of VR data learning analytics. In his book *Experience on Demand*, Jeremy Bailenson, who is also a cofounder of Strivr, describes how Strivr used data analytics to measure how quickly NFL players improved as they used VR. The player who improved the most was the Arizona Cardinals quarterback Carson Palmer, who happened to have a career-best year when he started using VR training. Bailenson speculates that Strivr and other systems like it "will be able to collect and analyze the massive amounts of data gathered by players like

Palmer, who has shown himself to be especially adept at teaching himself, and discover the best ways to implement the training."

It is likely that there will emerge positive applications for the data. One great example for the use of VR data is in the use of VR for racial bias training. Talespin—whose "firing Barry" training module we discussed in chapter 2—is one of the many start-ups offering VR as a solution for implicit bias and discrimination in the workplace. Cofounder Kyle Jackson described in a VentureBeat interview how the capacity to combine data analytics from gaze and verbal interactions with controlled changes to the virtual scenario opens up "all sorts of additional performance conversations" that weren't possible beforehand.[15] With the company's sales training modules, concrete differences were apparent in how trainees tried to sell services to an older male "virtual human" CEO versus a younger woman of color, putting their implicit bias on display in a way that might be able to improve the effectiveness of this kind of antidiscrimination training. How trainees spoke to the virtual agent, where they looked, and what they chose to say disclosed biases that they didn't realize they had.

A further example is grounded in a 2014 study conducted in Bailenson's Virtual Human Interaction Lab at Stanford, which analyzed nonverbal data during one-on-one student-teacher interactions.[16] In this study, led by Andrea Won, they found that they could predict students' test scores based on the analysis of gesture and posture data collected from two Microsoft Kinect sensors. Bailenson subsequently proposes that a VR system could therefore know how effectively a student is learning, in real time, and "transform and adjust on-the-fly."[17] But opportunities like these aren't all that's driving commercial investments in VR.

VR development is largely motivated by what sociologists Marion Fourcade and Kieran Healey call the *data imperative*—that is, the widespread institutional imperative to collect and analyze information about individuals' habits, tastes, values, and worth. As they write, "Contemporary organizations are both culturally impelled by the data imperative and powerfully equipped with new tools to enact it."[18] The array of sensors built into VR devices today render VR an increasingly powerful medium to enact this imperative.

Digital advertising is one powerful motivation for the commercial investment in VR, which makes possible the next step beyond common advertising identifiers, like the cookie. Internet cookies enable websites to track us between individual pages, remembering what we've added to our shopping cart or keeping us securely logged into our online banking sessions. Cookies are also a crucial element of the advertising infrastructure that enables targeted advertisements based on your internet viewing history. If I know you're reading about desk ergonomics, you're theoretically a good target for my advertisement for a standing desk. The challenge for internet advertisers is that tracking technologies like cookies can be deleted and blocked,[19] and the data we have about a web user is always incomplete. So new data points to bolster the digital advertising industry have emerged, including geolocation data yielded through a phone's GPS usage and through *mobile advertising identifiers* (MAIDs), device IDs that give app developers the ability to track users outside of their own apps. As digital media is in constant transition, so is the digital advertising industry always looking for new sources of data, and new ways that data can be operationalized in order to inspire confidence (and investment) in the industry.

It is in this context that we see the appeal of VR for a company like Facebook, which earns 97.9 percent of its revenue—over $80 billion a year—from advertising.

What makes VR so enticing—particularly to companies like Meta— is that its data is so intimate that it is a *behavioral biometric*.[20] As defined by the European Union's General Data Protection Regulation (GDPR), biometric data is "personal data resulting from specific technical processing relating to the physical, physiological, or behavioral characteristics of a natural person, which allows or confirms the unique identification of that natural person, such as facial images or dactyloscopic [fingerprint] data."[21]

In 2020, Mark Miller and colleagues published the results of their study that found that just five minutes of VR data—collected from watching a 360-degree video with an HTC VIVE headset—could be correctly relinked to a unique user with 95.3 percent accuracy.[22] Other researchers have confirmed these results, finding that it was possible to reidentify users even across different VR systems.[23]

Meta's data privacy policies and user license agreements for Oculus products currently offer limited insight into the company's plans for using VR data, but because it is extremely open-ended in what data collection and processing capacities it affords to Meta, it's easy to imagine the potential. In December 2019, Facebook informed customers that it would be using Oculus purchase data, but without regulation, it likely won't be long before Meta and others begin using the data collected from our use of VR to empower its powerful digital advertising arm to deliver targeted advertisements. The analysis of this data will likely reveal deep cognitive insights that go beyond what can currently be discerned from our internet browsing habits.

As Mark Pesce—the inventor of the sourceless orientation sensor—writes in the context of augmented reality, this capacity to "continuously profile its users, using the full array of sensory inputs, including gaze detection, developing a precisely detailed map of their engagement" will allow Meta to "know both the user's reaction to the real world environment, and any changes produced by those synthetic additions—information that could then be used to modify those additions to make them more engaging."[24] As with research into eye gaze and strategic thinking, how we look and interact with the world offers insights into how we think, our values and our attitudes, without us even realizing we're giving this up.

If this data is so effective, the potential implications for misuse and harm are enormous. Scandals such as the Facebook–Cambridge Analytica data scandal highlight the risks of personal data—in that case, quite basic information such as users' public profiles, page likes, ages, and locations—being used against the interests of the affected parties, including to deliver hypertargeted, inflammatory political advertising. Writing for the Electronic Frontier Foundation, Rory Mir and Katitza Rodriguez argue that "those in control of [VR] data may be able to identify patterns that let them more precisely predict (or cause) certain behavior and even emotions in the virtual world. It may allow companies to exploit users' emotional vulnerabilities through strategies that are difficult for the user to perceive and resist."[25] If VR becomes more widely used for work and education, it's difficult to understate the amount of data we will be giving up, about ourselves and the physical environments we use VR in, and even more difficult to speculate about how it might be used against us. "If privacy dies in VR," Mir and Rodriguez argue, then "it dies in real life."[26]

This belief in the accuracy and richness of VR data also leads to another risk as VR becomes more widely adopted and deployed. As a wide range of scholars note, data is not a neutral thing.[27] As Lisa Gitelman and Virginia Jackson put it, data is never raw: it's "cooked"—collected, stored, and circulated with specific aims and logics in mind. It's in the broader cultural misunderstanding about the objectiveness of data that many of VR's more immediate harms are likely to play out.

A FANTASY OF PERFECT DATA

In our research, we set out to study how it was that VR companies were using VR data and how they framed its potential. One of the most prominent things we saw across the companies we studied was the tendency to emphasize—particularly to institutional partners—how *perfect* this data is, particularly in the context of employee training. Unlike web-based tracking data, which is plainly incomplete, this VR data wasn't just good because of its capacity for immersive simulation, but because a VR system seamlessly collects a frameless data picture. For instance, Talespin's Kyle Jackson praises the fact that "we can measure anything, from your sentiment to your gaze to what you said and how you said it." Immerse cofounder Justin Parry describes VR as "fundamentally different" from other learning mediums, because they can "record absolutely everything that user did."

The idea that VR data "captures every detail," as Immerse claims, is based not just on the volume of data that VR creates but also on the idea that VR is psychologically real. As we discussed in chapter 2, the way that VR can create experiences that are so realistic that we react to them as though they were real is grounded in research like the use of VR by psychologists to treat arachnophobia.[28] Research like that is the basis for the approach by Marcus Carter (this book's coauthor) to use VR to connect zoo visitors more closely with animals.[29] The potential for VR as an assessment tool is also grounded in this idea: Strivr's Science Resources webpage, for instance, claims that VR simulations "activate the same neural pathways in the brain" as a real scenario would. This fantasy of assessment provides a rhetorical force behind the types of claims that companies—both developers and clients—are making about the quality, veracity, accuracy, and predictive power of their analytics. The overconfidence in this fantasy holds enormous potential for data-borne harm from VR.

Take, for instance, Strivr's partnership with Walmart. With forty-five training modules and seventeen thousand headsets in 4,700 locations, Strivr's VR tools let Walmart simulate events that would be difficult to run as physical training scenarios (such as a Black Friday shopping crowd), learn how to use new technology before it is installed, and perform soft skills training such as customer service tasks and dealing with difficult conversations with employees. These examples make sense as an application of VR as an educational technology in a business enterprise. There are also clear logistical benefits: an infrastructure of VR headsets means training can be quickly created and distributed across its stores at scale, circumventing the labor and travel costs and limitations of human teaching staff.

But Walmart isn't just using VR for training. Walmart is also working with Strivr to apply data analytics to these training scenarios and make decisions on the basis of these analytics. For example, in a blog post, Senior Vice President for Associate Experience Drew Holler describes how Walmart is using VR in the hiring process: "Rachel and a team of technology and business leaders developed a skills-based assessment that uses virtual reality to simulate everyday obstacles. Once a candidate completes a 15-minute assessment, leaders use the results to help them remove subjectivity and unconscious bias from the selection process. This solution enables a people-led, tech-empowered way of working."[30]

In related news coverage, Holler is quoted as saying the VR training is as important as knowledge of store departments, decision-making, leadership capacities, and soft skills. He describes an example in which the promotion—and a 10 percent pay raise—for a twelve-year employee was based in part on performance in the VR training. While Holler emphasizes that VR assessment is only one of the "data points" used in hiring decisions, many VR companies enthusiastically frame the potential for complete automation of these decisions.

For instance, one of Strivr's key patents is an "algorithm to predict how performance in a virtual environment will map to performance in that same situation or task in real life. This method automatically clusters learners into groups based on sensing data, which can include head, hand, and eye movements, as well as physiological data."[31] In a webinar, Michael Casale (chief science officer) says that the data Strivr collects (in this case, decision-making, performance, attention, and engagement) predict "in almost eighty percent or more than eighty percent of the cases how people would actually perform in the real world," subsequently suggesting that based on as little as twenty minutes of VR companies can "actually start to make predictions of real-world performance just based on what's going on in the headset, and that's of course incredibly efficient." Jeremy Bailenson—a Strivr cofounder—similarly proposes in his book *Experience on Demand* that a VR system could award grades based on the "continuous measure of learning and engagement formed from the analysis of literally millions of data points spanning hours per week for months at a time." Strivr is not alone in this: Immerse also suggests in a press release that VR has the potential to be used in recruitment, and in a press interview Mursion claims that the data gathered through its existing (non-AI-based) VR training will be able to automatically "measure human behavioral change." Murison even describes a project that aims to design performance tasks that "will be used as an alternative teacher certification assessment" in higher education.

The point we want to make here isn't that performance in VR can't sometimes predict success in the real world. It may actually help reduce bias in those decisions. But incorporating automated decision-making into VR on the back of a fantasy of perfect data is plagued with the potential for harm.

The example of automated decision-making that we are all perhaps most familiar with is Facebook's social media news feed that automatically determines—based on the view Meta has of our clicks, views, location data, and internet habits—what news we encounter on a daily basis. In his book *Automated Media*, Mark Andrejevic makes the point that personalized news is not necessarily bad, but in practice—where it has "arrived on the back of commercially owned and controlled platforms to service their advertising and marketing imperatives"[32]—it has served to exacerbate political divides,[33] provide a feeding ground for conspiracy theories and QAnon alternate realities,[34] and degrade democratic systems in the service of more time spent on Facebook's platform. Andrejevic's point here is that "the choice to implement automation within the existing socio-economic context carries with it a set of built-in tendencies that have important societal consequences."[35]

While advantageous to VR tech companies, it is crucial to remember that VR, and its data, is not the original experience. It is not an objective representation of the real thing. It is an interpretation of an interpretation, a fabrication. VR's fantasy of perfect data—that it captures for objective analysis a mirror-like reflection of the learning experience—is likely based on normative and exclusionary assumptions, which have in the past contained hidden gender, class, and racial biases.[36] For example, while we found no discussion of the training datasets in our research, commercial machine-learning products are typically trained on biased datasets of neurotypical, male, and able-bodied engineers.[37] In the context of VR learning analytics, this has the potential to codify xenophobia, ableism, and white supremacy within the black box of algorithmic bias, while avoiding critique because of the pervasive belief in the neutrality of data.[38] While Bailenson's idea for a VR system that can automatically assign grades based on VR data might work for some students, it is crucial to remember that implementing it runs the risk of misclassifying students whose gestures and postures fall outside the training data.

Strivr, for instance, describes in a webinar how it uses "verbal analytics" to measure "verbal fluency," which it suggests provides "objective" and "automated" predictions of a trainee's capability to deal with an emotional customer. Our concern is that the implementation of these technologies may overlook the limitations of speech recognition approaches, which "work best for white, highly educated, upper-middle class Americans,"[39] with recent research suggesting error rates almost twice as high for African Americans speakers versus white speakers (35 percent vs. 19 percent).[40] Presumably, age, physical fitness, and experience with VR technologies also all play a role—hidden in the "objectivity" of VR's embodiment data—in the predictive power of VR data analytics. On this basis, it is quite likely that without critically interrogating the VR fantasy of "perfect" data, the growing use of VR data analytics has the potential to exacerbate, rather than solve, issues of bias and discrimination.

The datafication of workplace decision-making (particularly to do with hiring and promotion) is an area that has been fraught with debate and critique.[41] As is now well established in the burgeoning field of AI ethics, transparency needs to be at the forefront of VR's development and deployment. This requires transparency into the data sets that are used for training VR analytics and into how algorithms make decisions and recommendations based on VR data, and the constant reminder that the data that VR generates is not perfect. It is, like all data, subjective and incomplete.

VIRTUAL REALITY DATA EMPIRES

In an episode of the *Georgian Impact Podcast*, Strivr founder and CEO Derek Belch discussed the expansiveness, and the value, of the data that the company has collected from its users so far:

Then when you think about machine learning and AI and all of this, well data begets data, it becomes this infinite loop . . . the more people that are using this training, the more data that we're collecting, the more data we have to build the models, the more data the models have, we have more people. It just circles itself. Interesting parallel for Strivr is self-driving cars . . . we have probably a hundred to a thousand times more data than anybody else. And so our models will be that much further along when they start to become more refined and more specific.

In this interview, Belch is pointing to a new educational data gold rush. Because Strivr owns the aggregate data associated with usage of its VR training tools and collects data about how this VR data maps to real world performance, Belch further suggests the potential for comparing employees across companies for the development of learning models and adaptive exams. A data platform engineer role advertised at Strivr claims that "tens of millions" of in-headset sessions will feed into "a streaming analytics platform that will allow us to process, join, aggregate, reform and query these very large structured and unstructured datasets to produce immersive analytics with deep insights on learning sessions." Interconnected with the issues we previously discussed around the fantasy of VR data, Belch goes on to envision a future in which these adaptive and autonomous simulations send users "down certain [learning] paths based on how you perform, and that's going to impact your score." Both the training and outcomes may become subject to future forms of analysis, segmentation, and bias.[42]

In short, Belch is claiming that whichever VR company acquires the most data first will get an unassailable lead over their competitors. That's why in the interview quoted at the beginning of this section, Belch compares the VR situation to the race to release a self-driving car, often measured by which technology company has collected the most data (enabling it to build the most accurate computer vision systems for navigation). Immerse echoes this speculation. In a webinar on data capture, Immerse's COO Justin Parry describes how VR experiences can be replayed, and therefore reanalyzed as analysis methods improve. All of this, of course, hinges on speculation that machine learning, data science, and artificial intelligence will find ways to refine these lakes of data and incorporate them into the development and sale of new products, which can similarly extract more data.

What this indicates is—perhaps unsurprisingly—the further expansion of data capitalism into VR technologies, as expounded by scholars such as Jathan Sadowski and Shoshanna Zuboff, who show that one of the key economic tendencies of big tech platforms is their extraction of data from large user bases. Yet as we have seen, data does not necessarily always neatly convert into monetary value. In fact, it is often accumulated but left unused.[43]

The typical history that is told about virtual reality is that the downfall of VR in the 1990s was due to the limitations of technology at the time. As VPL's Jaron Lanier puts it, VR was in a "waiting room for Moore's Law," which predicted exponential growth in computer capacity.[44] Certainly, in the 1990s, there was no computer that could create simulations as detailed as a VR headset in 2021. But VR's resurgence in the 2010s wasn't just because the technology had caught up or because Palmer Luckey captured the imagination of gamers with his Kickstarter campaign. The technological capacity had been there much earlier. The VR we have today is possible because of venture speculation in Silicon Valley and financial investment in dominating the VR market for the unending and unparalleled flow of data that it might create. Ultimately, then, VR is a product of data capitalism and is heading toward becoming one of its greatest contributors.

The potential for virtual reality to transform how we learn, play, and connect is profound, and it is rooted in VR's capacity to create simulations that feel *real*. But inextricably baked into this capacity is an advanced technological system of data collection. Each generation of VR technology has increased in its capacity to accurately surveil us and, now, the world around us. While this data collection makes VR possible and may offer new abilities and experiences that could make education more effective, games more engaging, and work tasks easier, the potential for misuse and harm is difficult to exaggerate. Can these added efficiencies ever outweigh the potential for data violence through algorithmic bias, discrimination, and surveillance? How can we approach a technology of such imbalanced potential?

7

OUR FANTASY

Through this book, we've highlighted the ways that VR—as a technology of modeling the world—is underlain by certain ways of seeing and understanding the world. Models of the world and ways of knowing—scientific, mathematical, and computational—are in fact overburdened by power relations, values, and social and political ideologies.[1] Writing at the time of the first commercial VR craze and the rise of commercial computing in the 1990s, feminist science and technology studies (STS) offered rich resources for pushing back against the notion of virtuality and VR's incorporeal nature, arguing that the virtual can never be disentangled from the politics of the social—with particular attention to gender and race.[2] As Nicola Green adroitly puts it, to "become virtual" is "not simply to use a computing system as a tool, nor is it to access a wholly 'other' space and become digital. Rather, it is a process of making connections between programmed and nonprogrammed spaces in specific locales, and power-laden social, cultural, and economic relationships."[3]

This is just as true today. As we've identified across the previous chapters of this book, VR cannot be disentangled from the politics of the material and ideological assumptions, logics, and values of those invested in its development and use. In interrogating these material and ideological assumptions, we've shown that for all the promise of a new technology that will radically disrupt the status quo, the solutions brought about by VR seem to intensify many of the same problems that exist in the present. By identifying the issues that exist in imagined futures of VR, we hope to offer a path for anticipating, addressing, and preventing the potential challenges of this technology before it becomes entrenched.

The prominence of the material basis of VR emerged clearly in chapter 2 through the deeply embodied phenomenon of VR-induced motion sickness. VR is a wearable technology that requires an intimate connection with the body to function correctly. When this connection fails—when there is a perception-proprioceptive conflict—motion sickness is the result, which shaped the failures of VR in the 1990s. As we charted, however, the politics of VR is a politics of some bodies and not others. The history of the development of VR saw it situated as an aspirational innovation in the service of hardcore gaming, and thus in service to the masculine gamer identity. The politics of this identity are embedded into the design and experience of VR, ultimately discriminating against women's bodies—across physical hardware, software, and online cultures—and people with disabilities. The origins of these discriminations are not just in the overrepresentation of men in VR development, but the libertarian, identity-free, and disembodied fantasies that underpin these developers' visions for VR. As we subsequently argued, the cost of this focus on the gamer body (at the expense of others) has been narrow and a self-reinforcing limit on the potential of the emerging medium for play.

Chapter 3 unpacked the politics of VR as a "disembodying" technology. Technosolutionist movements like VR for Good sought to expand the market for VR beyond gaming, represented best by the rapid uptake of the fantasy of VR as an "ultimate empathy machine." This emerged as a dominant use of VR, and a key way in which VR was evangelized—because the idea that VR could *solve* an intractable problem like empathy was specifically attractive to the types of people involved in VR discourse, funding, and production at the time. VR is not, of course, an empathy machine, but this fantasy is built upon the idea that VR's heightened presence and framelessness permits the (typically white, male, Silicon Valley entrepreneur) body to see through the eyes of another (typically brown, oppressed, vulnerable) body, an augmented perception that proponents claim inevitably leads to empathy and that ultimately relies on claims about vision, about empathy, and about the framelessness of VR that are false. What VR does accomplish is providing those (white, male, Silicon Valley entrepreneur) bodies a false proximity to the experiences of other people, a hyper-real spectacle that—in its illusion of disembodiment—becomes a barrier to actual change and empowerment. As feminist STS so clearly established in the 1990s, VR does not and cannot disembody the user. Our bodies—and their politics and privileges—are how we experience virtual realities, and only in acknowledging this can we unlock the actual benefits of the technology.

The acquisition of Oculus by Meta in 2014 placed modern VR on a trajectory toward the metaverse, which we approached in chapter 4 as a "fantasy of enclosure." For many of Meta's executives and engineers—such as Zuckerberg, Abrash, and Bosworth—the company's investment in the software and computing infrastructure for VR (and spatial computing beyond) is framed in terms of *perceptual enclosure*—creating environments that mediate our sense perceptions in ways that feel good, environments in which we might want to spend our lives. Yet the enclosure that the company imagines is more in line with Meta's current business model: a fantasy of mediating everyday life through a software and hardware stack fully controlled by Meta. Here, we focus on VR as material, in a sense familiar to those versed in Marxist thought—a lens through which to interrogate how the economic base of society (here Meta, one of the largest technology companies in the world) has real and palpable implications for how people think, feel, and act; the kinds of cultural and technical goods produced; and so on. We charted through our studies of Meta how the company is investing billions of dollars in building and normalizing the infrastructure that could enable it to commodify every aspect of our daily lives, a technological ambition underpinned by acquisitions and gargantuan investments in VR research and development. Through Meta, Zuckerberg is trying to stabilize—in the eyes of the public, regulators, and the company's shareholders—decades of competing discourses about what VR could be (the next paradigm of the internet, where more meaningful aspects of daily life are undertaken) and the potential for his company to totally control it.

Like chapter 2, chapter 5 called out the militarized and disciplinary institutions in which VR has its roots—and in which it remains in use today. Perhaps nowhere is this more concrete than in the fantasies of VR coordinating and training police and military in violence and the control of Black and brown people, where the outcomes can be extremely harmful (if not fatal). In charting the use of VR in the context of policing and military use, chapter 5 highlights how one of the dominant fantasies of

VR—that it can simulate real-world phenomena accurately—cannot be disentangled from realities of violence, particularly violence exerted by the police and military, institutions that (as we discussed) have deeply interwoven histories with VR technologies.

Finally, in chapter 6, we revisited the question of VR data—which underpins Meta's fantasy of enclosure—through an account of how VR companies are using VR data today. What we uncovered is a fantasy about quantifying bodies for analysis and instrumentation, a fantasy that perpetuates many of the same harms of algorithmic bias and discrimination in the use of AI that are now well-known and widely criticized. Once again, this fantasy is one that will most likely benefit some people and harm others—most likely those whose bodies do not match the socially constructed idea of the "normate" body, codified by a homogenous and financially motivated development process. The point of this chapter was not that VR data can't offer insights, but that the fantasy of "perfect" data that VR boosters perpetuate will likely lead to more harm because of the confidence that decision-makers will place in decisions made based on this data. When coupling this potential for harm with visions of an increasingly expansive, quotidian metaverse, chapter 6 forewarns all of us about the potential for further expansions of algorithmic control.

VR is a technology that is sometimes good, but it is also one that can be misdirected, bad, and even directly harmful. By being attentive to the fantasies that are dominant around VR today, we—academics, policymakers, consumers—are better situated to call out the bad and harmful ones, and to redirect the future of VR to one that is beneficial for the many rather than the few.

Fantasies are a critical site for making such an intervention. As Jathan Sadowski writes, and as is clear in our account of the future-leaning discourses of VR boosters, "Capital seeks to assert dominion over *the future*—constraining what type of social change is viable."[4] Yet for the most part, these futures haven't yet materialized; futures exist on a largely discursive register.[5] As such, we still have the opportunity to shape and resist these visions—looking beyond their often cynical (or harmful) outlooks. Put differently, fantasies are *performative*; they structure and legitimate technology development, but they are not inevitable. We can imagine futures for VR play that offer more inclusive experiences, appealing to a broader variety of audiences than just gamers. We can demand proof that VR for good actually effects positive change and does not just give the appearance of action. We can demand principles of openness, anonymity, and privacy before VR expands further, before it becomes entrenched, and before harms have taken effect.

We began developing the proposal for this book in late 2020, and submitted it to the publisher after the generous advice of our editor, David Weinberger, shortly before Marcus had a baby in mid-2021. While the proposal was in review, in a single day, Zuckerberg and his Meta rebrand fundamentally and totally redefined popular discourses about VR, an era-defining event for our area of scholarship. In June 2023, Apple emerged as an entrant into the VR (or more specifically, mixed reality) market with its own novel product category—*spatial computing* (resisting the branding of the metaverse)—suggesting that its proprietary computing platform will define areas like work and entertainment.

As practitioners of sociotechnical research, to bear witness to these shifts has been an interesting, unsettling, and at times confounding experience. Prior to the metaverse—and to a lesser extent, Apple's spatial computing intervention—we at times felt adrift, with relatively few others, in a small boat, futilely attempting to gain the attention of the much larger collection of boats circling the topics of AI and big data. For us, VR entering into popular and academic lexica was as though a big tech aircraft carrier (or perhaps, luxury super yacht) had sailed at full speed through our small patch of water, throwing us overboard and dragging in its wake a full complement of tech boosters, crypto bros, and intellectual hot takes. Suddenly the spotlight was on the topic, but the attention frustratingly remained on the wrong things, and in a way that ignored the many present and immediate harms of VR. It has decidedly not been an easy intellectual environment in which to write a book on the topic.

We initially grappled with revising the structure of the book and incorporating the concept of the metaverse more centrally. After all, in early 2022, it seemed like the metaverse had fully broken through popular discourse and would continue to define the field for sufficiently long that a book on virtual reality would feel immediately out of date when published. But as we turned our attention to this concept, pulled back the curtain on all the hype and investment, we found very little worth discussing: an empty room, quite a lot like *Second Life*, where nobody has any legs, with someone trying to sell you a JPEG. What we concluded was that the term *metaverse* was (and is) acting as a floating signifier, similar to crypto buzzwords like *decentralization*—which, as Nathan Schneider notes, have a functional lack of clarity that "enables people of varying ideological persuasions to imagine themselves as part of a common project."[6] As with crypto, this lack of clarity means that critique of the metaverse cannot mean the dismissal of the imprecise and anticipatory vision. The metaverse is a powerful rhetorical device precisely because of this. Zuckerberg often frames the metaverse as a decade-long project, an act of futuring that puts substantive discussion today out of sight, and clouds over present-day issues as only temporary. As such, we reversed course, and the book today remains largely unchanged from what we initially proposed before the metaverse, a critical interrogation of the fantasies that underpin the VR technologies that exist today.

Being adjacent to this rebrand and subsequent popular interest has been fascinating. One immediate effect has been the rapid proliferation of academic work on the metaverse. Much of this work is opportunistic, an expression of the pressures placed on academics in increasingly neoliberalized universities to publish fast, and on topics that receive citations and impact quickly. For instance, national funding schemes such as the National Science Foundation or Australian Research Council (ARC) become vehicles for boosting the concept of the metaverse by funding research that seeks to make it possible (two projects recently funded by the ARC base their case on the metaverse), because uncritical reproduction of Meta's claims in a grant application boost the case for a research project's potential for impact and importance. For funders, the metaverse is new ground they haven't funded before, encouraging academics to legitimate the topic further in the design and focus of their research. The danger is equally present for critical scholarship. As STS scholar Alfred Nordmann articulates in the context of nanotechnologies, even critical tech research can reproduce and even increase the hype surrounding emerging technologies, lending legitimacy to industry claims "simply by taking them seriously." [7] Most urgently, critical tech studies of the metaverse—as easy as it is to critique—can also act as a distraction. The effect of this scholarship is the further legitimization of the tech industry's interests.

Yet high-profile cases tied to the metaverse and spatial computing have also helped galvanize critical work on VR. In our experience, increasing attention being paid to VR has drawn us to other researchers with common political or normative commitments in studying the unique cultural and ethical challenges posed by VR. The increasing prominence of VR within tech sector visions of the computing future has reinforced the need now—more than ever—for what we've elsewhere dubbed *critical VR studies*[8]—critical intellectual work seeking to understand the implications of VR on society and economy across material-technical, discursive, and political economic registers.

Increasing attention to spatial computing and the metaverse have also catalyzed conversations about governance—such as how regulation and policy might be applied to VR. Although there are issues in VR—identified in this book—that align with regulatory concerns (from data privacy to antitrust), calls for regulatory oversight can be problematic when we are working in such speculative terrain. Saying something needs to be regulated makes it seem more credible. The WEF's metaverse initiative (one that has received funding from Meta), for instance, legitimates not only the inevitability of the metaverse but also ideas like "no single company will own or dominate the metaverse" (a quote taken directly from the front page of the WEF initiative). This parrots a common line delivered by Meta's VP for global affairs, Nick Clegg, in interviews, that "no one company will own and operate the metaverse." Claims like these—legitimated by "independent" organizations like the WEF—shape what kinds of regulatory conversations can happen about a technology.

Paradoxically, we'd also argue that what we've seen of the attention of regulatory campaigners on regulating the metaverse actually has the effect of making it a *less urgent* regulatory initiative because, in comparison to emerging technologies of the day like generative AI, the framing of the metaverse means it is unseen and undefined. It is a distraction from regulating what needs to be regulated today. What we've shown through this book is that we do not need to regulate the metaverse; we do not need VR laws, just more robust tech laws in general. We need laws that protect the use of technology—VR included—against people's interests.

We need stronger laws that prohibit unnecessary and carceral control and surveillance, in workplaces and society writ large. We need protections from using technology in decision-making without evidence. These are not unique to VR, but emblematic of the broader failures of technology regulation we're facing today.

Concepts like spatial computing and the metaverse are futuristic, imprecisely defined, and ultimately hollow ideals. They should not distract us from the fact that VR, a technology with a long history, is currently happening and is absolutely here to stay. In its current form, the fantasies that surround VR have the potential to cause harm, and its further proliferation and expansion will only exacerbate these harms. It is our intent that through calling out these fantasies, we will move closer to a future that doesn't perpetuate but instead eliminates these inequalities.

ACKNOWLEDGMENTS

We begin by acknowledging, and paying respect to, the traditional owners of the land on which we wrote this book: the Gadigal people of the Eora Nation, and the Turrbal and Yugara people.

We are deeply grateful to the individuals and institutions who have played an integral role in the development of our research into virtual reality and the completion of this book. With sincerest apologies to those who we have omitted while rushing to finish this acknowledgements section to meet a deadline, we would specifically like to thank the following: Heather Horst and Gerard Goggin, whose support for this project and mentorship at its inception was a critical vote of confidence; our colleagues and collaborators in XR, Kate Clark, Luke Heemsbergen, Joanne Gray, Mark Pesce, and John Tonkin; our vigilant proofreader, Premeet Sidhu; at QUT, Ben's mentor, Amanda Lotz, and the QUT Game Studies Feedback group—Brendan Keogh, Benjamin Nicoll, Erin Maclean, and Dan Padua; at the University of Melbourne, Marcus's VR-in-the-zoo collaborators Sarah Webber and Wally Smith; at Zoos Victoria, David Methven, Liz Liddicoat, Sally Sherwen, and Emily Mcleod; and at PHORIA, Joseph Purdam and Trent Clews-de Castella.

We'd also like to thank our editors at the MIT Press, Gita Manaktala and David Weinberger, whose advice and feedback on our proposal put us on a path with this manuscript that has been incredibly rewarding to follow.

On a personal note, Marcus thanks his loving wife, Vicki, whose patience in listening to him talk about VR is unparalleled, and his son, Tommy, whose curiosity and joy in all things has been a life-changing inspiration and a highly valued distraction from academic work. Ben would like to express his gratitude to his partner, Jade, for her love and support throughout the book-writing process (and, in particular, her vigilant efforts in helping him not overwork).

NOTES

CHAPTER 1

1. Nieborg and Helmond, "Political Economy of Facebook's Platformization."

2. Marcus, *Technoscientific Imaginaries.*

3. See, e.g., Bijker, Hughes, and Pinch, *Social Constructions of Technological Systems.*

4. Berlant, *Cruel Optimism.*

CHAPTER 2

1. Vinciguerra, "Tom Kalinske Talks."

2. Hecht, "Optical Dreams, Virtual Reality."

3. Characterizing the development of VR at the time, Alex Smith—lead programmer on one of the other games (*Outlaw Racing*)—"never saw the prototype hardware," so it's unknown how many games were even implemented with support for Sega VR. Via Whitehouse, "Sega VR Reviewed."

4. Whitehouse, "Sega VR Reviewed."

5. Virtuality went insolvent in 1997, having struggled to find commercial success with coin-op arcades. See McFerran, "Reality Crumbles."

6. Woolley, *Virtual Worlds.*

7. Without head tracking, and with a typical console controller, Steven Boyer describes the Virtual Boy as "essentially a console version of previous attempts at 3-D goggle peripherals." A reasonable argument could be made that the NES game *Duck Hunt*—which used an innovative light-gun gadget, rather than a controller—is a closer precursor to the incorporation of the body via natural interfaces of contemporary VR than the Virtual Boy is. Boyer, "Virtual Failure," 27.

8. See Morrissette, "How Games Marketing Invented Toxic Gamer Culture"; Lien, "No Girls Allowed"; and Kocurek, *Coin-Operated Americans.*

9. Boyer, "Virtual Failure," 27.

10. Boyer, 27.

11. See https://www.mtbs3d.com for archived MTBS3D forum posts. In this book, we do not provide full references for quoted text from sources such as forums,

YouTube comments, and VR company websites since permanent links to content are rarely available.

12. Kumparak, "Brief History of Oculus."

13. Robertson and Zelenko, "Voices from a Virtual Past."

14. Ewalt, *Defying Reality*, 95.

15. Foxman, "Making the Virtual a Reality."

16. This subsequently embroiled Oculus, Luckey, Carmack, and Facebook in a high-profile litigation with ZeniMax (which Carmack was working for at the time, and which owned Doom 3 publisher Bethesda/id Software). ZeniMax alleged intellectual property theft. After being initially awarded $500 million in damages, this was reduced to $250 million on appeal (although the full settlement remains confidential).

17. Welsh, "John Carmack."

18. Welsh.

19. Luckey, "Update #36."

20. Evans, *Re-emergence of Virtual Reality*.

21. See "Oculus Rift: Step into the Game," Kickstarter, https://www.kickstarter.com /projects/1523379957/oculus-rift-step-into-the-game.

22. "Oculus Rift: Step into the Game."

23. Core values that are intrinsically tied up with masculine notions of power and with traditionally masculine genres like "action, sports racing, shooting, and fighting games." See Cote, *Gaming Sexism*, 28.

24. Slater et al., "Immersion, Presence and Performance in Virtual Environments," 165.

25. Ermi and Mäyrä, "Fundamental Components of the Gameplay Experience," 8.

26. Ermi and Mäyrä, 8.

27. Although $2 billion is the typically quoted purchase price for Oculus, the subsequent lawsuit between John Carmack's previous employer, ZeniMax, and Oculus/ Facebook revealed that the purchase price was $3 billion total, including additional retention and milestone bonuses. See Gilbert, "Facebook Just Settled."

28. In his book *Experience on Demand*, VR researcher Jeremey Bailenson, who runs the Virtual Human Interaction Lab at Stanford University, describes a visit Zuckerberg made to his lab shortly before the Oculus acquisition closed. Also see Woputz, "Giving Mark Zuckerberg a Demo."

29. Foxman, "Making the Virtual a Reality," 99.

30. David Ewalt, "Minecraft Creator Kills Oculus Rift Plans."

31. Drawing from the German philosopher Martin Heidegger, Liel Leibovitz argues that interactive software—in his case, video game—artifacts outside the human mind powerfully shape cognition and embodied action. For Heidegger, technological objects (and their specific material affordances) have the possibility for enframing

(*ge-stell*), which, to put it very simply, sets the parameters of how the world is seen. Crucially, for Heidegger's account of technology and Leibovitz's account of video games, the way that this happens is often intuitive (rather than something noticeable). Leibovitz, *God in the Machine*. For other researchers in the phenomenological (and, specifically, Heideggerian) tradition, studying software like video games, the aim is to follow and update Heidegger's aim of understanding how *specific* material properties of technological objects shape human action and cognition (see, e.g., James Ash's account of elements of game interfaces in *The Interface Envelope*). While we do not draw from Heidegger—or any specific phenomenological tradition—in this book, much of the thinking in this chapter is broadly influenced by this general perspective that media-specific affordances shape user experience in particular ways.

32. Munafo, Diedrick, and Stoffregen, "Virtual Reality Head-Mounted Display."

33. Golding, "Far from Paradise," 341.

34. Golding, 341.

35. Harley, "Palmer Luckey," 1154.

36. Foxman, "Making the Virtual a Reality," 99.

37. Munafo, Diedrick, and Stoffregen, "Virtual Reality Head-Mounted Display."

38. See danah boyd, "Is the Oculus Rift Sexist?"

39. See boyd.

40. Peck, Sockol, and Hancock, "Mind the Gap."

41. On this latter example, Robertson points out the impact of having women involved in VR development: "One eye-tracking headset stubbornly ignored my pupils until an employee asked if I was wearing mascara. When it got recalibrated perfectly a few minutes later, I was surprised—not by the fact that it worked, but by the fact that anyone had thought to troubleshoot makeup. Incidentally, this was one of the only VR startups I've ever covered with a female founder." See Robertson, "Building for Virtual Reality?"

42. Robertson.

43. Via Heaney, "Data Suggests." The (Anthropometric Survey) ANSUR II is a public data set created by the US Army in 2012, based on measurements of US army soldiers and reservists.

44. Heaney, "Data Suggests."

45. Stanney, Fidopiastis, and Foster, "Virtual Reality Is Sexist."

46. Stanney et al.'s study included measurements one hour post exposure, so this was not trivial motion sickness.

47. Belamire, "My First Virtual Reality Groping."

48. Outlaw, "Virtual Harassment."

49. Belamire, "My First Virtual Reality Groping."

50. Dibbell, "Rape in Cyberspace."

51. Jackson and Schenker, "Dealing with Harassment in VR."

52. Center for Countering Digital Hate, "Horizon Worlds Exposed."

53. *Metaverse: Another Cesspool of Toxic Content.*

54. Parshall, "Assault on Your VR Body."

55. Maloney, Freeman, and Robb, "Virtual Space for All."

56. Maloney, Freeman, and Robb, "It Is Complicated."

57. Clark and Le, "Sexual Assault in the Metaverse."

58. Danaher, "Law and Ethics of Virtual Sexual Assault."

59. This is inextricably an act of nonconsensual sexual dominance: *teabagging* is a slang term for a sexual act.

60. As a further example, we'd be remiss here not to mention *Dead or Alive Xtreme 3*, a VR game available for the PlayStation 4 that allows players to grope the game's bikini-clad nonplayer characters while they grimace, protects their bodies with their arms, and say, "I don't like it." In the video demo shared online that precipitated media attention, the male-sounding audience laughs in response. As legal scholar Mary Franks writes, "The primary concern with games like these is not the harm one user inflicts on another actual user in a virtual reality environment, but the harmful habits the technology encourages the user to indulge." Franks, "Desert of the Unreal," 528. Writing for Engadget, Sean Buckley describes *Dead or Alive Xtreme 3* as "sexual assault, the game." Buckley, "'Dead or Alive.'"

61. Via the scholarship of disability studies scholar Rosemary Garland-Thomson, *Extraordinary Bodies.*

62. Kilteni, Groten, and Slater, "Sense of Embodiment in Virtual Reality."

63. As a starting point for this rich field of research, see Goggin and Newel, *Digital Disability*; Ellis and Goggin, *Disability and the Media*; Ellcessor, *Restricted Access*; and Costanza-Chock, *Design Justice.*

64. Paterson, "On Haptic Media."

65. Gerling and Spiel, "Critical Examination of Virtual Reality."

66. AbleGamers Charity, "Virtual Reality."

67. https://communityforums.atmeta.com/t5/Get-Help/Utter-Disappointment/m-p/864722.

68. Harley, "Palmer Luckey."

69. If you ever want to frustrate a VR enthusiast, or VR techwriter, ask them this question. Our apologies for using it rhetorically here.

70. In contrast, in less time, Sony has shipped thirty million PS5 systems, even while they were difficult to purchase for most of that time. The total number of Quest devices shipped is also a little misleading; Meta is guarded about the number of *active* users of its headsets, and about how many devices are played with a few times at Christmas and then locked away in a cupboard. The highest figure we could locate is three million active users in the popular VR gaming and social application *Rec Room*, far short of the 110-million-plus monthly active users on PlayStation Network.

71. Boluk and Lemieux, *Metagaming*.

72. Mobile VR headsets (such as Meta's Quest line) are all-in-one devices, so the computing power is significantly reduced in comparison to a headset (such as the HTC VIVE) that runs off of a desktop computer.

73. This way of conceptualizing the form of immersion in *Super Hot VR* can be understood through Grant Bollmer and Adam Suddarth's concept of *embodied parallelism*. Bollmer and Suddarth argue that immersion in VR is premised on these explicit engagements between the body and the virtual environment. Bollmer and Suddarth, "Embodied Parallelism and Immersion."

74. Valve, "Best of 2022—Best of VR."

75. Kocurek, *Coin-Operated Americans*.

76. Morrissette, "How Games Marketing Invented Toxic Gamer Culture."

77. Eklund, "Who Are the Casual Gamers?"

78. Chess, "Hardcore Failure in a Casual World," 60–61.

79. Consalvo, "Hardcore Casual."

80. Consalvo and Paul, *Real Games*.

81. There is a breadth of scholarship examining #GamerGate, detailing its right-wing politics, racist and sexist underpinnings, and real-world harms. As a starting point, see Butt and Apperley, "Vivian James"; Chess and Shaw, "Conspiracy of Fishes"; Golding and Van Deventer, *Game Changers*; and Cote, *Gaming Sexism*.

82. Foxman, "Making the Virtual a Reality," 99.

83. Savage, "Inside Valve."

84. Kuchera, "Superhot VR."

85. Bezmalinovic, "Beat Saber."

86. Simon, *Wii Are Out of Control*.

87. As we put the finishing touches on this book, rumors have begun circulating that Nintendo is working with Google on a VR headset. As Marcus argues elsewhere with Tianyi Zhangshao, Nintendo is a pioneer not just in the "casual" market with the Nintendo Wii, but more recently in hybridizing "hardcore" and "casual" play with the Nintendo Switch, essentially collapsing the (already flawed) distinction. Our expectation is that Nintendo may be the one company whose unique approach to innovation in gaming experiences might deliver a widely successful consumer VR headset. See Zhangshao and Carter, "Hybrid Revolution."

CHAPTER 3

1. Victorian-era zoos saw captive animals housed in orderly, adjacent cages, structured by a popular fascination with natural history and the developing field of zoology. Underpinned by principles of scientific observation, enclosures were designed to maximize visibility to visitors, and animals had no choice or control over their visibility. Animal welfare was subsequently extremely poor. In contrast, the modern

zoo is motivated—and morally justified—by principles of conservation education and animal welfare.

2. This approach is often expressed in the literature as an attempt to "illustrate to guests what life in the wild might be like for the animals on view" and is understood to help visitors better connect the relationship that wild animals must have with their natural environment, and thus the importance of protecting that natural environment to protect the animal. See Kutska, "Variation in Visitor Perceptions."

3. Thomas, *Social Change for Conservation.*

4. In our work, we make the argument that these kinds of digital augmentations will empower zoos to design enclosures in ways that facilitate greater animal welfare outcomes, because digital technologies will be able to pick up the slack, so to speak. See the work described in Carter, Sherwen, and Webber, "Evaluation of Interactive Projections"; Webber et al., "Interactive Technology and Human-Animal Encounters at the Zoo"; Dong et al., "Supporting Animal Welfare"; and Webber et al., "Kinecting with Orangutans."

5. Carter et al. "Virtual Reality in the Zoo."

6. See https://www.ted.com/talks/chris_milk_how_virtual_reality_can_create_the _ultimate_empathy_machine?language=en.

7. Oculus VR, "VR forGood."

8. "Roger Ebert on Empathy."

9. Heft-Luthy, "Myth of the 'Empathy Machine.'"

10. Rose, "Immersive Turn."

11. United Nations SDG Action Campaign, "Syrian Refugee Crisis."

12. Milk, "How Virtual Reality Can Create."

13. Bollmer, "Empathy Machines," 63.

14. Bailenson, *Experience on Demand.*

15. Zuckerberg, "I'm excited to announce that we've agreed to acquire Oculus VR, the leader in virtual reality technology," Facebook, March 25, 2014.

16. Mark Zuckerberg, "Founders Letter, 2021."

17. Hassapopoulou, "Playing with History," 369.

18. Hassapopoulou, 369. See also Nakamura, "Feeling Good about Feeling Bad."

19. Rose, "Immersive Turn."

20. Milk, "How Virtual Reality Can Create."

21. Vaselli, "Changing Humanitarianism for the Better?"

22. Zuckerberg, "I'm excited . . ."

23. Irom, "Virtual Reality and the Syrian Refugee Camps," 4287.

24. Irom, 4283.

25. Irom, 4287.

26. Chouliaraki, "'Improper Distance,'" 368.

27. Chouliaraki, 365.

28. Nash, "Virtual Reality Witness," 125.

29. Nash, 125.

30. Here we see Paul Bloom's case, represented in his book *Against Empathy*, for one of the many limits of empathy as an emotion and motivator for moral action.

31. In her critique, Lisa Nakamura goes further, situating forms of "antiracist" VR films alongside meal replacement packs and other Silicon Valley technologies created to automate and "hack" bodily labors. Here, "VR automates the Labor of feeling . . . what is being cured is not the body's need for food, but rather society's [in particular, privileged white Silicon Valley engineers'] need to feel better about itself by feeling something about racism." Nakamura, "Feeling Good about Feeling Bad," 60.

32. Nakamura, "Feeling Good about Feeling Bad," 125.

33. Solon, "Mark Zuckerberg 'Tours' Flooded Puerto Rico."

34. As Joshua Fisher and Sarah Schoemann aptly put it, "The immediate corporate commodification of pain trespassed many ethical boundaries." Fisher and Schoemann, "Toward an Ethics of Interactive Storytelling."

35. Urry, "'Consumption' of Tourism."

36. Hassapopolou, "Playing with History," 372; emphasis in original.

37. See https://www.facebook.com/zuck/videos/10104094186863501/?comment_id =180060679227397&reply_comment_id=627147064341885&comment_tracking=% 7B%22tn%22%3A%22R9%22%7D.

38. Harley, "Virtual Bodies Inc.," 254.

39. Jerrett, Howell, and Dansey, "Developing an Empathy Spectrum for Games." For an analysis of popular discourse around empathy in VR, see also Foxman, Markowitz, and Davis, "Defining Empathy."

40. Andrejevik and Voclic, "Virtual Empathy," 300

41. Ellis, *Documentary: Witness and Self-Revelation*.

42. Bollmer, "Empathy Machines."

43. In 2021, BuzzFeed reported that Mursion employed white actors to roleplay as black employees, such as for a roleplaying scenario in which a (white) manager roleplayed as an empathetic manager to a black employee (played by a white actor). In the reportage, Apryl Williams—a race and digital media scholar from the University of Michigan—argues that "you can't separate this from the history of blackface, yellowface, and redface in this country, even if you have the most sensitive actors in the world playing these characters." See Baker-White, "'This Is Blackface.'"

44. Anand and Winters, "Retrospective View of Corporate Diversity Training."

45. Dobbin and Kalev, "Why Sexual Harassment Programs Backfire."

46. Anand and Winters, "Retrospective View of Corporate Diversity Training."

47. Dobbin and Kalev, "Promise and Peril of Sexual Harassment Programs."

48. Dobbin and Kalev, "Why Sexual Harassment Programs Backfire."

49. Robb and Doverspike, "Self-Reported Proclivity to Harass."

50. US Equal Employment Opportunity Commission, "Task Force Co-chairs Call on Employers and Others."

51. Kaplan et al., "Effects of Virtual Reality, Augmented Reality, and Mixed Reality"; and Jensen and Konradsen, "Review of the Use of Virtual Reality Head-Mounted Displays."

52. Rivero, "Consulting Firms Are Betting on Unproven VR Diversity Trainings."

53. This refers to direct responses, such as confronting the harasser and telling them to stop, and formally reporting harassment to an authority figure. Rawski and colleagues speculate that this result could arise because "participants in the VR condition did not perceive the harassment as severe enough to report," which could reflect a difference in how (reenacted) sexual harassment is perceived in the 2D video versus the "realistic" video. Alternatively, another possible explanation provided was that "participants immersed in VR may have had more fears about the social-psychological-emotional consequences of reporting," which the authors generously suggest provides "some evidence that participants were deeply, emotionally engaged in the practice scenario." Rawski, Foster, and Bailenson, "Sexual Harassment Bystander Training Effectiveness."

54. Janet Murray, "Not a Film and Not an Empathy Machine."

55. Reuters, "You're Fired!"

56. Bergman, "Barry's Not Sobbing."

57. Goleman, *Emotional Intelligence*.

58. Heft-Luthy, "Myth of the 'Empathy Machine.'"

59. The FTC sued Meta in July 2022, filing an injunction to block this acquisition on the ground that it "may yield multiple harmful outcomes, including less innovation, lower quality, higher prices, less incentive to attract and keep employees, and less consumer choice." In February 2023, a judge ruled that the purchase could proceed. FTC v. Meta Platforms, Inc. [07/27/22] N.D. Cal, https://archive.org/details/gov.uscourts.cand.398508/gov.uscourts.cand.398508.1.0.pdf.

CHAPTER 4

1. O'Dell, "Facebook Is Buying Oculus."

2. Meta, "First Quarter 2022 Results Conference Call."

3. Jaron Lanier, 2017, cited in Evans, *Re-emergence of Virtual Reality*, 31.

4. As Jaron Lanier points out in his book *Dawn of the New Everything*, the failure of VPL was specifically attributed to the inability of the company to arrive on a unified direction for it to take—with it split between experimentation and the sale of low-cost, high-margin commercial products.

5. Mark Zuckerberg, "I'm excited . . . ."

6. Rubin, "Inside Story of Oculus Rift."

7. Hoffmann, Proferes, and Zimmer, "'Making the World More Open and Connected."

8. This conference ran until it was merged with the Facebook/Meta Connect conference in 2020.

9. Meta Quest, "Michael Abrash Keynote from F8 2015."

10. Heath, "This Is Meta's AR/VR Hardware Roadmap through 2027."

11. Meta Quest, "Oculus Connect Keynote: Michael Abrash.

12. Meta Quest.

13. Daymond, "Mark Zuckerberg Talks Future."

14. Cited in Egliston and Carter, "Oculus Imaginaries."

15. Meta Quest, "Day 1 Keynote."

16. Meta Quest.

17. Cited in Egliston and Carter, "'Metaverse and How We'll Build It.'"

18. UploadVR, "Andrew Bosworth OC6."

19. UploadVR, "Michael Abrash OC6."

20. Meta Quest, "Oculus Connect 5."

21. Meta Quest, "Horizon Worlds."

22. Kennedy et al., *Digital Domesticity*, 57.

23. Basu, "Metaverse Has a Groping Problem Already."

24. Gillespie, *Custodians of the Internet.*

25. Murphy, "How Will Facebook Keep Its Metaverse Safe for Users?"

26. Murphy.

27. Reality Labs, "From the Lab to the Living Room."

28. Saker and Frith, "Coextensive Space," 1426.

29. Meta Quest, "Oculus Connect 4."

30. Levy, "Metaverse Is Mark Zuckerberg's Mobile Do-Over."

31. Bundeskartellamt, "Meta (Facebook) Responds to the Bundeskartellamt's Concerns."

32. This approach has not been without critique. See Harley, "Promise of Beginnings."

33. Gorbett, "Announcing Our XR Hackathon Winners."

34. Gorbett.

35. Cited in Oremus, "In 2021, Tech Talked up 'the Metaverse.'"

36. Egliston and Carter, "'Metaverse and How We'll Build It.'"

37. Barnett, "Convening Publics," 411.

38. Zuckerberg, "2015 06 22 Mark's Vision."

39. Oculus VR, "Introducing App Lab."

40. Robertson, "Facebook Is Turning VR into a Platform."

41. Partin, "Bit by (Twitch) Bit," 1.

42. Nix and Gurman, "Meta's Oculus Unit."

43. Trimananda et al., "OVRseen."

44. Egliston and Carter, "Examining Visions of Surveillance."

45. Meta Quest, "Michael Abrash Keynote from F8 2015."

46. Meta Quest, "Michael Abrash Keynote from F8 2015."

47. Matney, "Oculus Research Is Now the 'Facebook Reality Lab.'"

48. Hesch, Kozminski, and Linde, "Powered by AI."

49. Although Meta had made VR-related acquisitions prior to the formation of Reality Labs, as we have shown elsewhere, the Reality Labs period represents the highest volume of VR-related acquisitions (and acquisitions more generally) in the company's entire history. See Egliston and Carter, "'Metaverse and How We'll Build It.'"

50. Federal Trade Commission v. Meta Platforms Inc., 2022 US District Court, San Jose, 5:22-cv-04325-EJD (2022).

51. Silberling, "FTC Puts Zuckerberg on the Stand."

52. Reality Labs, "Inside Facebook Reality Labs."

53. Hesch, Kozminski, and Linde, "Powered by AI."

54. Nix and Gurman, "Meta's Oculus Unit"; Silberling, "FTC Puts Zuckerberg on the Stand"; European Parliament, "Metaverse: Opportunities, Risk, and Policy Implications"; and eSafety Commissioner, "Immersive Technologies—Position Statement."

55. Táíwò, *Elite Capture*.

56. Notably, Facebook suggests that a principles-based approach to regulation (following from its RI principles) will be key in the legislative regulation of VR. For example, in a response to a review of the Australian Privacy Act published in December 2020, Facebook noted of VR regulation that a "prescriptive set of rules may deter innovation and result in a net negative outcome for consumers" and that its staff instead "strongly support the retention of a flexible and principles-based approach to privacy regulation."

57. Applin and Flick, "Facebook's Project Aria."

58. Hallinan, Brubaker, and Fiesler, "Unexpected Expectations."

59. To be completely transparent, in 2022, we were awarded funding for a project titled Disability and Accessibility in the Metaverse from Meta Australia.

60. See de Hoop, Pols, and Romijn, "Limits to Responsible Innovation," 118.

61. Stilgoe, Owen, and Macnaghten, "Developing a Rramework for Responsible Innovation."

62. Phan et al., "Economies of Virtue."

63. Clegg, "Making the Metaverse."

64. Clegg.

65. Abrash, "VR's Grand Challenge."

66. Oremus, "In 2021, Tech Talked up 'the Metaverse.'"

67. Ball, "Metaverse: What It Is."

68. Lai, "Meet Me in the Metaverse."

69. See Sadowski and Beegle, "Expansive and Extractive Networks of Web3."

70. Meta, "First Quarter 2022 Results Conference Call."

71. Daymond, "Mark Zuckerberg Talks Future."

72. Daymond.

73. Robinson, "Sci-Fi Guru Who Predicted Google Earth."

74. Karpf, "Virtual Reality is the Rich White Kid of Technology."

75. Zitron, "Mark Zuckerberg Is a Liar."

76. Roose, "Metaverse Is Mark Zuckerberg's Escape Hatch."

77. Seufert, "I dislike using 'metaverse' . . . ."

78. Pahwa, "Why the Metaverse Has to Look So Stupid."

79. Reality Labs, "Project Aria: Facebook Connect."

80. Applin and Flick, "Facebook's Project Aria," 1.

81. Tarnoff, *Internet for the People.*

82. Matsuda, "Hyper-Reality."

83. Rodriguez, "Facebook's Meta Mission."

84. Alex Heath, "This Is Meta's AR/VR Hardware Roadmap."

CHAPTER 5

1. Carter, "West Coast Police Adopt Virtual Reality Training Simulators."

2. Haraway, "Situated Knowledges," 581.

3. Winner, "Do Artifacts Have Politics?"

4. Galtung, "Cultural Violence," 291.

5. Geoghegan, "Ecology of Operations," 61.

6. For an excellent overview of Sutherland's role in the militarized origins of VR, see Harley, ""This Would Be Sweet in VR.'"

7. Images from Sutherland, "Head-Mounted Three-Dimensional Display."

8. Computer History Museum, "Ivan Sutherland."

9. Computer History Museum.

10. SIGGRAPH Conferences, "VR @ 50."

11. Birt and Furness, "Visually Coupled Systems."

12. Birt and Furness.

13. Birt and Furness, 29.

14. Aitoro, "30 Years."

15. Ewalt, *Defying Reality*.

16. Ewalt.

17. For an overview of Goertz's influence on haptics, see Parisi, *Archaeologies of Touch*, 220–229.

18. United States Army Acquisition Support Center, "Synthetic Training Environment (STE)."

19. Gervais, "Synthetic Training Environment Revolutionizes Sustainment Training."

20. IITSEC, "Army Training Mission at I/ITSEC 2021."

21. À la Paul Scharre's conceptualisation of decentralized—data-driven and networked—warfare as a "swarm." See Scharre, "How Swarming Will Change Warfare."

22. Suits, "Synthetic Training Environment to Enhance Soldier Lethality."

23. Raytheon Intelligence and Space, "Synthetic Training Environment."

24. Kratos Defense & Security Solutions, "Simulation Training Platform."

25. Applied Virtual Simulation, "Protected Mobility Tactical Trainer."

26. Saab, "The secondary trigger . . . ."

27. Novet, "Microsoft Wins U.S. Army Contract."

28. Cited in Egliston and Carter, "'Interface of the Future.'"

29. Geoghegan, "Ecology of Operations," 86

30. Geoghegan, 88.

31. Congressional Research Service, *Advanced Battle Management System (ABMS)*.

32. See IQGeo's website at https://www.iqgeo.com.

33. In-Q-Tel, "Immersive Wisdom Secures Strategic Investment."

34. Immersive Wisdom, "Immersive Wisdom Awarded."

35. Immersive Wisdom, "Remote Ops Center."

36. Chamayou, *Theory of the Drone*.

37. The company's name, Anduril, is inspired by Aragorn's sword Andúril, from the *Lord of the Rings* books.

38. Hollister, "BlueShark."

39. Simonite, "Behind Anduril's Effort."

40. Simonite. A more recent update to Anduril's website reads: "The Lattice User Interface (UI) elegantly allows users to visualize and interact with thousands of sensors and effectors in a single UI. Operators can seamlessly scale from tactical to

strategic views. The Lattice UI spans web, desktop, mobile, and VR modalities to fit your need." Anduril, "Lattice."

41. White House, "Fact Sheet."

42. Anduril, "Lattice."

43. Beaumont, "'Never Sleeps, Never Even Blinks.'"

44. Crogan, "War, Mathematics, and Simulation," 655.

45. It is particularly important to note that these visions of the military adoption of VR are not incidental. These are discourses that have been taken up in the context of attempts to think about VR within the context of international policymaking about war and counterterrorism, such as in a 2021 panel convened by the United Nations Office of Counterterrorism. The goal of this panel was to raise awareness of how VR technologies might militate against threats of terrorism and violent extremism, rather than to focus on the harms inherent within these technologies and the institutions that employ them for purposes of counterterrorism. Booster discourses—about VR's capacity to mediate military decision-making through the simulated rendition of information—are taken up uncritically into policymaking conversations about their utility.

46. Virilio and Lotringer, *Pure War.*

47. Dyer-Witherford and de Peuter, *Games of Empire*, 100–101.

48. Graham, *Cities under Siege*, xiv.

49. Egliston and Carter, "'Interface of the Future.'"

50. Carter, "West Coast Police Adopt Virtual Reality Training Simulators"; Open Innovation Team, "Investigating Virtual Reality for the College of Policing"; and Lam, "VR Group Lands $1.7m WA Police Deal."

51. Axon, "Never Miss a Moment."

52. Axon, "Solutions for Federal Civilian and Defense."

53. Axon, "Axon VR Training."

54. Axon Inc.

55. DeGeurin, "Police VR Training."

56. See https://jigsaw.google.com/.

57. See https://jigsaw.google.com/trainer/.

58. Jigsaw, "Immersive Technology to Help Advance Public Safety."

59. McQuillan, *Resisting AI*, 22.

60. Axon, "Axon and Fūsus Partner." See also Axon, "VR Simulator Training Is Here."

61. Dowse et al., *Research Report.*

62. Wang, *Carceral Capitalism.*

63. Hodge, System and method for personalized virtual reality experience.

64. Cox, "Prison Company Patents VR to Give Inmates Brief Taste of Freedom."

CHAPTER 6

1. See Rheingold, *Virtual Reality*.

2. Barlow, "Being in Nothingness."

3. Silverman, "NASA Explores Virtual Reality."

4. Chung et al., "Exploring Virtual Worlds with Head-Mounted Displays."

5. Zimmerman et al., "Hand Gesture Interface Device." See also US Patent 4,542,291.

6. NASA Spinoff, "Computer Technology: A New Continent of Ideas."

7. See US Patent 5,526,022-A.

8. Personal communication with Mark Pesce, email, March 2021.

9. Robertson and Zelenko, "Voices from a Virtual Past."

10. Orland, "Rift Goes Wireless."

11. Hesch, Kozminski, and Linde, "Powered by AI."

12. Han et al., "Using Deep Neural Networks."

13. Bailenson, "Protecting Nonverbal Data Tracked in Virtual Reality."

14. Newn et al., "Looks Can Be Deceiving."

15. Takahashi, "How VR Can Help."

16. Won, Bailenson, and Janssen, "Automatic Detection of Nonverbal Behavior."

17. Bailenson, *Experience on Demand*, 249.

18. Fourcade and Healey, "Seeing Like a Market."

19. In fact, it is estimated that around 60 percent of digital advertisements are not even seen by human eyes. See Hwang, *Subprime Attention Crisis*.

20. Pfeuffer et al., "Behavioural Biometrics in VR."

21. See https://gdpr-info.eu/art-4-gdpr/.

22. Miller, Banerjee, and Banerjee, "Within-System and Cross-System."

23. Miller et al., "Personal Identifiability of User Tracking Data."

24. Pesce, *Augmented Reality*.

25. Mir and Rodriguez, "If Privacy Dies in VR."

26. Mir and Rodriguez.

27. Kitchin, *Data Revolution*.

28. Suso-Ribera et al., "Virtual Reality, Augmented Reality, and *In Vivo* Exposure Therapy."

29. Carter et al., "Virtual Reality in the Zoo."

30. Holler, "Customers Are Changing."

31. See US Patent 10,586,469-B2. Quote is from a 2020 Strivr press release.

32. Andrejevic, *Automated Media*, 38.

33. Statt, "Facebook Reportedly Ignored Its Own Research."

34. Warzel, "Is QAnon the Most Dangerous Conspiracy Theory?"

35. Andrejevic, *Automated Media*, 26.

36. See Noble, *Algorithms of Oppression*; Crawford, *Atlas of AI*; and Hoffmann, "Terms of Inclusion."

37. Mehrabi et al., "Survey on Bias and Fairness in Machine Learning."

38. Shea Swauger, in "Our Bodies Encoded," makes this argument in the context of automated exam proctoring in higher education.

39. Harwell, "Accent Gap."

40. Koenecke et al., "Racial Disparities in Automated Speech Recognition."

41. Kim, "Data-Driven Discrimination at Work,"; Rosenblat, *Uberland*; Raghavan et al., "Mitigating Bias in Algorithmic Hiring"; and Sánchez-Monedero, Dencik, and Edwards, "What Does It Mean?"

42. As discussed in Amoore, *Cloud Ethics*.

43. Sadowski, "When Data Is Capital."

44. Evans, *Re-emergence of Virtual Reality*.

CHAPTER 7

1. This is what feminist philosopher of science and technology Donna Haraway would call *situated knowledge*. As Haraway writes, claims to objectivity or neutrality in science are fallacious. They do not offer impartiality, a "view from above, from nowhere." See Haraway, "Situated Knowledges," 589.

2. For instance, see Balsamo, *Technologies of the Gendered Body*; Sophia, "Virtual Corporeality: A Feminist View"; and Green, "Disrupting the Field."

3. Green, "Disrupting the Field."

4. Sadowski, "Future Schlock"; emphasis in original.

5. Bryant and Knight, *Anthropology of the Future*.

6. Schneider, "Decentralization: An Incomplete Ambition."

7. Nordmann, "If and Then."

8. Carter and Egliston, "What Are the Risks of Virtual Reality Data?," 486.

BIBLIOGRAPHY

AbleGamers Charity. "Virtual Reality: New Technology, New Challenges." *AbleGamers Blog*, March 8, 2017. https://ablegamers.org/virtual-reality-new-technology-new -challenges/.

Abrash, Michael. "VR's Grand Challenge: Michael Abrash on the Future of Human Interaction." *Meta Quest Blog*, Meta, July 24, 2017. https://www.oculus.com/blog /vrs-grand-challenge-michael-abrash-on-the-future-of-human-interaction/.

Aitoro, Jill. "30 Years: Virtual Reality—Training Transformation." *Defense News*, October 25, 2016. https://www.defensenews.com/30th-annivesary/2016/10/25/30-years -virtual-reality-training-transformation/.

Amoore, Louise. *Cloud Ethics: Algorithms and the Attributes of Ourselves and Others.* Durham, NC: Duke University Press, 2020.

Anand, Rohini, and Mary-Frances Winters. "A Retrospective View of Corporate Diversity Training from 1964 to the Present." *Academy of Management Learning & Education* 7, no. 3 (2008): 356–372. https://doi.org/10.5465/amle.2008.34251673.

Andrejevic, Mark. *Automated Media.* Oxfordshire, UK: Routledge, 2020.

Andrejevic, Mark, and Zala Voclic. "Virtual Empathy." *Communication, Culture & Critique* 13, no. 3 (2020): 295–310. https://doi.org/10.1093/ccc/tcz035.

Anduril. "Lattice." Accessed March 23, 2023. https://www.anduril.com/lattice/.

Applied Virtual Simulation. "Protected Mobility Tactical Trainer: Scalable, Accessible and Effective Training for PMV Crews." Accessed March 23, 2023. https://applied virtual.net/pmtt.

Applin, Sally A., and Catherine Flick. "Facebook's Project Aria Indicates Problems for Responsible Innovation When Broadly Deploying AR and Other Pervasive Technology in the Commons." *Journal of Responsible Innovation* 5 (2021): 1–15. https://doi .org/10.1016/j.jrt.2021.100010.

Ash, James. *The Interface Envelope.* New York: Bloomsbury, 2015.

Axon. "Axon and Fūsus Partner to Provide Enhanced Real-Time Community Policing Solutions." Press release, May 19, 2022. https://investor.axon.com/2022-05-19-Axon-and-Fusus-Partner-to-Provide-Enhanced-Real-Time-Community-Policing-Solutions.

Axon. "Axon VR Training." Accessed March 23, 2023. https://www.axon.com/training/vr.

Axon. "Never Miss a Moment: How the Axon Ecosystem Can Help You Capture the Digital Evidence You Need." Axon, October 31, 2022. https://www.axon.com/news/technology/never-miss-a-moment-how-the-axon-ecosystem-can-help-you-capture-the-digital-evidence-you-need.

Axon. "Solutions for Federal Civilian and Defense." Accessed March 23, 2023. https://au.axon.com/industries/federal.

Axon. "VR Simulator Training Is Here—Now Available for Public Safety to Increase De-Escalation Training." Axon, May 24, 2022. https://www.axon.com/news/vr-simulator-training-is-here-now-available-for-public-safety-to-increase-de-escalation-training.

Bailenson, Jeremy. *Experience on Demand: What Virtual Reality Is, How It Works, and What It Can Do*. New York: W. W. Norton, 2018.

Bailenson, Jeremy. "Protecting Nonverbal Data Tracked in Virtual Reality." *JAMA Pediatrics* 172, no. 10 (2018): 905–906. https://doi.org/10.1001/jamapediatrics.2018.1909.

Baker-White, Emily. "'This Is Blackface': White Actors Are Playing Black Characters in Virtual Reality Diversity Training." BuzzFeed News, December 1, 2021. https://www.buzzfeednews.com/article/emilybakerwhite/diversity-training-mursion-vr-white-actors.

Ball, Matthew. "The Metaverse: What It Is, Where to Find It, and Who Will Build It." MatthewBall.co, January 13, 2020. https://www.matthewball.vc/all/themetaverse.

Balsamo, Anne. *Technologies of the Gendered Body: Reading Cyborg Women*. Durham, NC: Duke University Press, 1996.

Barlow, John Perry. "Being in Nothingness: Virtual Reality and the Pioneers of Cyberspace." *Mondo 2000*, no. 2 (Summer 1990): 34–44. https://archive.org/details/Mondo.2000.Issue.02.1990/mode/2up.

Barlow, John Perry. "Virtual Reality and the Pioneers of Cyberspace." *WIRED*, April 30, 2015. https://www.wired.com/2015/04/virtual-reality-and-the-pioneers-of-cyberspace.

Barnett, Clive. "Convening Publics: The Parasitical Spaces of Public Action." In *The SAGE Handbook of Political Geography*, edited by Kevin R. Cox, Murray Low, and Jennifer Robinson, 403–418. London: SAGE, 2008.

Basu, Tanya. "The Metaverse Has a Groping Problem Already." *MIT Technology Review*, December 16, 2021. https://www.technologyreview.com/2021/12/16/1042516/the -metaverse-has-a-groping-problem/.

Beaumont, Hilary. "'Never Sleeps, Never Even Blinks': The Hi-Tech Anduril Towers Spreading Over the US Border." *Guardian*, September 16, 2022. https://www .theguardian.com/us-news/2022/sep/16/anduril-towers-surveillance-us-mexico -border-migrants.

Belamire, Jordan. "My First Virtual Reality Groping." Medium, October 20, 2016. https://medium.com/athena-talks/my-first-virtual-reality-sexual-assault-2330410 b62ee.

Bergman, Ben. "Barry's Not Sobbing: VR Training Platform Talespin Triples Funding." dot.LA, March 3, 2020. https://dot.la/talespin-virtual-reality-training-26453 61687.html.

Berlant, Lauren. *Cruel Optimism*. Durham, NC: Duke University Press, 2011.

Bezmalinovic, Tomislav. "Beat Saber: The Early Days of the Most Successful VR Game of All Time." MIXED, June 18, 2022. https://mixed-news.com/en/beat-saber -the-beginnings-of-the-most-successful-vr-game-of-all-time/.

Bijker, Wiebe E., Thomas P. Hughes, and Trevor Pinch, eds. *The Social Constructions of Technological Systems*. Cambridge, MA: MIT Press, 1987.

Birt, Joseph A., and Thomas A. Furness. "Visually Coupled Systems." *Air University Review* 25, no. 3 (March/April 1974): 28–40. https://www.airuniversity.af.edu/Portals /10/ASPJ/journals/1974_Vol25_No1-6/1974_Vol25_No3.pdf.

Bloom, Paul. *Against Empathy: The Case for Rational Compassion*. New York: Penguin Random House, 2017.

Bollmer, Grant. "Empathy Machines." *Media International Australia* 165, no. 1 (2017): 63–76. https://doi.org/10.1177/1329878X17726794.

Bollmer, Grant, and Adam Suddarth. "Embodied Parallelism and Immersion in Virtual Reality Gaming." *Convergence: The International Journal of Research into New Media Technologies* 28, no. 2 (2022): 579–594. https://doi.org/10.1177/13548565211070691.

Boluk, Stephanie and Patrick Lemieux. *Metagaming*. Minneapolis: University of Minnesota Press, 2017.

boyd, danah. "Is the Oculus Rift Sexist?" Quartz, March 28, 2014. https://qz.com /192874/is-the-oculus-rift-designed-to-be-sexist.

Boyer, Steven. "A Virtual Failure: Evaluating the Success of Nintendo's Virtual Boy." *Velvet Light Trap* 64 (2009): 23–33. https://doi.org/10.1353/vlt.0.0039.

Bryant, Rebecca, and Daniel Knight. *The Anthropology of the Future*. Cambridge: Cambridge University Press, 2019.

Buckley, Sean. "'Dead or Alive' VR Is Basically Sexual Assault, the Game." Engadget, last updated August 29, 2016. https://www.engadget.com/2016-08-29-dead-or-alive -vr-is-basically-sexual-assault-the-game.html.

Bundeskartellamt. "Meta (Facebook) Responds to the Bundeskartellamt's Concerns—VR Headsets Can Now Be Used without a Facebook Account." Press release, November 23, 2022. https://www.bundeskartellamt.de/SharedDocs/Meldung/EN /Pressemitteilungen/2022/23_11_2022_Facebook_Oculus.html.

Butt, Mahli-Ann Rakkomkaew, and Thomas Apperley. "Vivian James—the Politics of #Gamergate's Avatar." In *Abstract Proceedings of the First International Joint Conference of DiGRA and FDG*, 1–6. Dundee, Scotland: Digital Games Research Association and Society for the Advancement of the Science of Digital Games, August 2016. http:// www.digra.org/wp-content/uploads/digital-library/paper_154.pdf.

Carter, John. "West Coast Police Adopt Virtual Reality Training Simulators." Apex Officer, October 19, 2022. https://www.apexofficer.com/resources/west-coast-police -adopt-virtual-reality-training-simulators.

Carter, Marcus, and Ben Egliston. "What Are the Risks of Virtual Reality Data? Learning Analytics, Algorithmic Bias and a Fantasy of Perfect Data." *New Media & Society* 25, no. 3 (2023): 485–504.

Carter, Marcus, Sally Sherwen, and Sarah Webber. "An Evaluation of Interactive Projections as Digital Enrichment for Orangutans." *Zoo Biology* 40, no. 2 (2021): 107–114. https://doi.org/10.1002/zoo.21587.

Carter, Marcus, Sarah Webber, Simon Rawson, Wally Smith, Joseph Purdam, and Emily McLeod. "Virtual Reality in the Zoo: A Qualitative Evaluation of a Stereoscopic Virtual Reality Video Encounter with Little Penguins (Eudyptula Minor)." *Journal of Zoo and Aquarium Research* 8, no. 4 (2020): 239–245. https://doi.org/10.19227/jzar .v8i4.500.

Center for Countering Digital Hate. "Horizon Worlds Exposed." Center for Countering Digital Hate, March 8, 2023. https://counterhate.com/research/horizon-worlds -exposed/.

Chamayou, Grégoire. *A Theory of the Drone*. New York: New Press, 2015.

Chess, Shira. "Hardcore Failure in a Casual World." *Velvet Light Trap* 81 (2018): 60– 62. https://muse.jhu.edu/pub/15/article/686904#sub02.

Chess, Shira, and Adrienne Shaw. "A Conspiracy of Fishes, or, How We Learned to Stop Worrying About #GamerGate and Embrace Hegemonic Masculinity." *Journal of Broadcasting & Electronic Media* 59, no. 1 (2015): 208–220. https://doi.org/10.1080 /08838151.2014.999917.

Chouliaraki, Lilie "'Improper Distance': Towards a Critical Account of Solidarity as Irony." *International Journal of Cultural Studies* 14, no 4. (2011): 363–381. https://doi .org/10.1177/1367877911403247.

Chung, James C., Mark R. Harris, Fredrick P. Brooks, Henry Fuchs, Michael T. Kelley, John Hughes, Ming Ouh-Young, Clement Cheung, Richard L. Holloway, and Michael Pique. "Exploring Virtual Worlds with Head-Mounted Displays." In *Three-Dimensional Visualization and Display Technologies* 1083 (1989): 42–52. https://apps .dtic.mil/sti/pdfs/ADA208088.pdf.

Clark, Kate Euphemia, and Trang Le. "Sexual Assault in the Metaverse Isn't a Glitch That Can Be Fixed." Lens, Monash University, October 13, 2022. https://lens .monash.edu/@politics-society/2022/10/13/1385033/sexual-assault-in-the-metaverse -isnt-a-glitch-that-can-be-fixed.

Clegg, Nick. "Making the Metaverse: What It Is, How It Will Be Built, and Why It Matters." Medium, May 18, 2022. https://nickclegg.medium.com/making-the -metaverse-what-it-is-how-it-will-be-built-and-why-it-matters-3710f7570b04.

Computer History Museum. "Ivan Sutherland: 'Virtual Reality Before It Had That Name.'" Computer History Museum, episode 15, February 12, 2019. https:// computerhistory.org/ivan-sutherland-virtual-reality-before-it-had-that-name-playlist/.

Congressional Research Service. *Advanced Battle Management System (ABMS)*. Washington, DC: Congressional Research Service, February 15, 2022. https://sgp.fas.org /crs/weapons/IF11866.pdf.

Consalvo, Mia. "Hardcore Casual: Game Culture Return(s) to Ravenhearst." In *Proceedings of the 4th International Conference on Foundations of Digital Games*, 50–54. New York: Association for Computing Machinery, 2009. https://doi.org/10.1145 /1536513.1536531.

Consalvo, Mia, and Christopher A. Paul. *Real Games: What's Legitimate and What's Not in Contemporary Videogames*. Cambridge, MA: MIT Press, 2019.

Costanza-Chock, Sasha. *Design Justice: Community-Led Practices to Build the Worlds We Need*. Cambridge, MA: MIT Press, 2020.

Cote, Amanda. *Gaming Sexism*. New York: New York University Press, 2020.

Counterpoint Research. *Global AR & VR Headsets Market Share: Q1 2021—Q4 2022*. Counterpoint Research, accessed March 2023. https://www.counterpointresearch .com/global-xr-ar-vr-headsets-market-share/.

Cox, Joseph. "Prison Company Patents VR to Give Inmates Brief Taste of Freedom." Vice Motherboard, September 9, 2021. https://www.vice.com/en/article/3aqm4k /prison-virtual-reality-vr-global-tel-link.

Crawford, Kate. *Atlas of AI*. New Haven: Yale University Press, 2021.

Crogan, Patrick. "War, Mathematics, and Simulation: Drones and (Losing) Control of Battlespace." In *Zones of Control: Perspectives on Wargaming*, edited by Pat Harrigan and Matthew G. Kirschenbaum, 641–668. Cambridge, MA: MIT Press, 2016.

Danaher, John. "The Law and Ethics of Virtual Sexual Assault." In *Research Handbook on the Law of Virtual and Augmented Reality*, edited by Woodrow Barfield and Marc Jonathan Blitz, 363–388. Cheltenham, UK: Edward Elgar Publishing, 2018.

Daymond, John. "Mark Zuckerberg Talks Future of the Metaverse with Daymond John." Recorded panel from SXSW 2022. YouTube, March 19, 2022. https://www .youtube.com/watch?v=El19o1Ib-HE&ab_channel=DaymondJohn.

DeGeurin, Mack. "Police VR Training: Empathy Machine or Expensive Distraction?" Gizmodo, May 25, 2022. https://gizmodo.com.au/2022/05/police-vr-training -empathy-machine-or-expensive-distraction/.

de Hoop, Evelien, Auke Pols, and Henny Romjin. "Limits to Responsible Innovation." *Journal of Responsible Innovation* 3, no. 2 (2016): 110–134. https://doi.org/10 .1080/23299460.2016.1231396.

Dibbell, Julian. "A Rape in Cyberspace." *Village Voice*, October 18, 2005. https:// www.villagevoice.com/2005/10/18/a-rape-in-cyberspace/.

Dobbin, Frank, and Alexandra Kalev. "The Promise and Peril of Sexual Harassment Programs." *Proceedings of the National Academy of Sciences* 116, no. 25 (2019): 12255– 12260. https://doi.org/10.1073/pnas.1818477116.

Dobbin, Frank, and Alexandra Kalev. "Why Sexual Harassment Programs Backfire." *Harvard Business Review*, May/June 2020. https://hbr.org/2020/05/why-sexual -harassment-programs-backfire.

Dong, Ruining, Marcus Carter, Wally Smith, Zaher Joukhadar, Sally Sherwen, and Alan Smith. "Supporting Animal Welfare with Automatic Tracking of Giraffes with Thermal Cameras." In *Proceedings of the 29th Australian Computer Human Interaction Conference (OZCHI '17)*, 386–391. New York: Association for Computing Machinery, 2017. https://doi.org/10.1145/3152771.3156142.

Dowse, Leanne, Simone Rowe, Eileen Baldry, and Michael Baker. *Research Report: Police Responses to People with Disability*. Royal Commission into Violence, Abuse, Neglect and Exploitation of People with Disability, October 2021. https://disability .royalcommission.gov.au/publications/police-responses-people-disability.

Dyer-Witherford, Nick, and Greig de Peuter. *Games of Empire: Global Capitalism and Video Games*. Minneapolis: University of Minnesota Press, 2009.

Egliston, Ben and Marcus Carter. "Examining Visions of Surveillance in Oculus' Data and Privacy Policies, 2014–2020." *Media International Australia* 188, no. 1 (2023): 52– 66. https://doi.org/10.1177/1329878X211041670.

Egliston, Ben, and Marcus Carter. "'The Interface of the Future': Mixed Reality, Intimate Data and Imagined Temporalities." *Big Data & Society* 9, no. 1 (2022): 1–15. https://doi.org/10.1177/20539517211063689.

Egliston, Ben, and Marcus Carter. "'The Metaverse and How We'll Build It': The Political Economy of Meta's Reality Labs." *New Media & Society* (2022): 1–25. https:// doi.org/10.1177/14614448221119785.

Egliston, Ben, and Marcus Carter. "Oculus Imaginaries: The Promises and Perils of Facebook's Virtual Reality." *New Media & Society* 24, no. 1 (2022): 70–89. https://doi.org/10.1177/1461444820960411.

Eklund, Lina. "Who Are the Casual Gamers? Gender Tropes and Tokenism in Game Culture." In *Social, Casual and Mobile Games*, edited by Tama Leaver and Michelle Wilson, 15–30. London: Bloomsbury, 2015.

Ellcessor, Elizabeth. *Restricted Access: Media, Disability, and the Politics of Participation*. New York: New York University Press, 2016.

Ellis, John. *Documentary: Witness and Self-Revelation*. Oxfordshire, UK: Routledge, 2011.

Ellis, Katie, and Gerard Goggin. *Disability and the Media*. London: Palgrave Macmillan, 2015.

Ermi, Laura, and Frans Mäyrä. "Fundamental Components of the Gameplay Experience: Analysing Immersion." In *Proceedings of the 2005 DiGRA International Conference: Changing Views: Worlds in Play*, 1–14. Pittsburgh: ETC Press, 2005. http://www.digra.org/wp-content/uploads/digital-library/06276.41516.pdf.

eSafety Commissioner. "Immersive Technologies—Position Statement." eSafety Commissioner, December 10, 2020. https://www.esafety.gov.au/industry/tech-trends-and-challenges/immersive-tech.

European Parliament. "Metaverse: Opportunities, Risks, and Policy Implications." Think Tank, briefing, June 24, 2022. https://www.europarl.europa.eu/thinktank/en/document/EPRS_BRI(2022)733557.

Evans, Leighton. *The Re-emergence of Virtual Reality*. Oxfordshire, UK: Routledge, 2019.

Ewalt, David M. *Defying Reality: The Inside Story of the Virtual Reality Revolution*. New York: Penguin, 2018.

Ewalt, David M. "Minecraft Creator Kills Oculus Rift Plans Because Facebook Creeps Him Out." *Forbes*, March 25, 2014. https://www.forbes.com/sites/davidewalt/2014/03/25/minecraft-creator-kills-oculus-rift-plans-because-facebook-creeps-him-out/?sh=328ec2b018f3.

Fisher, Joshua A., and Sarah Schoemann. "Toward an Ethics of Interactive Storytelling at Dark Tourism Sites in Virtual Reality." In *Proceedings of Interactive Storytelling: 11th International Conference on Interactive Digital Storytelling*, 577–590. New York: Springer International Publishing, 2018.

Fourcade, Marion, and Kieran Healy. "Seeing Like a Market." *Socio-Economic Review* 15, no. 1 (2017): 9–29. https://doi.org/10.1093/ser/mww033.

Foxman, Maxwell. "Making the Virtual a Reality." *Digital Culture & Society* 7, no. 1 (2021): 91–110. https://doi.org/10.14361/dcs-2021-0107.

Foxman, Maxwell, David M. Markowitz, and Donna Z. Davis. "Defining Empathy: Interconnected Discourses of Virtual Reality's Prosocial Impact." *New Media & Society* 27, no. 8 (2021): 2167–2188. https://doi.org/10.1177/1461444821993120.

Franks, Mary Anne. "The Desert of the Unreal: Inequality in Virtual and Augmented Reality." *UC Davis Law Review* 51 (2017): 499–538.

Galtung, Johan. "Cultural Violence." *Journal of Peace Research* 27, no. 3 (1990): 291–305. https://doi.org/10.1177/0022343390027003005.

Geoghegan, Bernard Dionysius. "An Ecology of Operations: Vigilance, Radar, and the Birth of the Computer Screen." *Representations* 147, no. 1 (2019): 59–95. https://doi.org/10.1525/rep.2019.147.1.59.

Gerling, Kathrin, and Katta Spiel. "A Critical Examination of Virtual Reality Technology in the Context of the Minority Body." In *Proceedings of the 2021 CHI Conference on Human Factors in Computing Systems*, 1–14. New York: Association for Computing Machinery, 2021. https://doi.org/10.1145/3411764.3445196.

Gervais, Maria R. "The Synthetic Training Environment Revolutionizes Sustainment Training." US Army, August 23, 2018. https://www.army.mil/article/210105 /the_synthetic_training_environment_revolutionizes_sustainment_training.

Gilbert, Ben. "Facebook Just Settled a $500 Million Lawsuit over Virtual Reality after a Years-Long Battle—Here's What's Going On." *Business Insider*, December 12, 2018. https://www.businessinsider.com/facebook-zenimax-oculus-vr-lawsuit-explained -2017-2.

Gillespie, Tarleton. *Custodians of the Internet: Platforms, Content Moderation, and the Hidden Decisions That Shape Social Media*. New Haven, CT: Yale University Press, 2021.

Goggin, Gerard, and Christopher Newell. *Digital Disability: The Social Construction of Disability in New Media*. Washington, DC: Rowman & Littlefield, 2003.

Golding, Dan. "Far from Paradise: The Body, the Apparatus and the Image of Contemporary Virtual Reality." *Convergence: The International Journal of Research into New Media Technologies* 25, no. 2 (2019): 340–353. https://doi.org/10.1177/135485651 7738171.

Golding, Dan, and Leena Van Deventer. *Game Changers: From Minecraft to Misogyny, the Fight for the Future of Videogames*. Melbourne: Affirm Press, 2016.

Goleman, Daniel. *Emotional Intelligence: Why It Can Matter More Than IQ*. New York: Bantam Books, 1995.

Gorbett, Brian. "Announcing Our XR Hackathon Winners." *News For Developers* (blog), Meta, December 7, 2021. https://developers.facebook.com/blog/post/2021 /12/07/announcing-xr-hackathon-winners/.

Graham, Stephen. *Cities under Siege: The New Military Urbanism*. New York: Verso Books, 2010.

Gray, Kishonna L. "Solidarity Is for White Women in Gaming." In *Diversifying Barbie and Mortal Kombat: Intersectional Perspectives and Inclusive Designs in Gaming*, edited by Yasmin B. Kafai, Gabriela T. Richard, and Brendesha M. Tynes, 59–70. Pittsburgh: ETC Press, 2016.

Green, Nicola. "Disrupting the Field: Virtual Reality Technologies and 'Multisited' Ethnographic Methods." *American Behavioral Scientist* 43, no. 3 (1999): 409–421. https://doi.org/10.1177/00027649921955344.

Hallinan, Blake, Jed R. Brubaker, and Casey Fiesler. "Unexpected Expectations: Public Reaction to the Facebook Emotional Contagion Study." *New Media & Society* 22, no. 6 (2020): 1076–1094. https://doi.org/10.1177/1461444819876944.

Han, Shangchen, Beibei Liu, Tsz Ho Yu, Randi Cabezas, Peizhao Zhang, Peter Vajda, Eldad Isaac, and Robert Wang. "Using Deep Neural Networks for Accurate Hand-Tracking on Oculus Quest." *AI Meta* (blog), September 25, 2019. https://ai.facebook.com/blog/hand-tracking-deep-neural-networks.

Haraway, Donna. "Situated Knowledges: The Science Question in Feminism and the Privilege of Partial Perspective." *Feminist Studies* 14, no. 3 (1988): 575–599. https://doi.org/10.2307/3178066.

Harley, Daniel. "Palmer Luckey and the Rise of Contemporary Virtual Reality." *Convergence: The International Journal of Research into New Media Technologies* 26, no. 5–6 (2020): 1144–1158. https://doi.org/10.1177/1354856519860237.

Harley, Daniel. "The Promise of Beginnings: Unpacking 'Diversity' at Oculus VR." *Convergence: The International Journal of Research into New Media* (2022): 1–15. https://doi.org/10.1177/13548565221122911.

Harley, Daniel. "'This Would Be Sweet in VR': On the Discursive Newness of Virtual Reality." *New Media & Society* (2022): 1–17. https://doi.org/10.1177/14614448221084655.

Harley, Daniel. "Virtual Bodies Inc.: Framing Corporate Mediations of Bodies in VR." *Public* 30, no. 60 (2020): 250–259. https://doi.org/10.1386/public_00019_7.

Harwell, Drew. "The Accent Gap." *Washington Post*, July 19, 2018. https://www.washingtonpost.com/graphics/2018/business/alexa-does-not-understand-your-accent/.

Hassapopoulou, Marina. "Playing with History: Collective Memory, National Trauma, and Dark Tourism in Virtual Reality Docugames." *New Review of Film and Television Studies* 16, no. 4 (2018): 365–392. https://doi.org/10.1080/17400309.2018.1519207.

Heaney, David. "Data Suggests Oculus Rift S IPD Range 'Best' For Just Half of Adults." *UploadVR*, April 5, 2019. https://uploadvr.com/data-suggests-oculus-rift-s-ipd-range-best-for-around-half-of-adults/.

Heath, Alex. "This is Meta's AR/VR Hardware Roadmap through 2027." Verge, February 2018, 2023. https://www.theverge.com/2023/2/28/23619730/meta-vr-oculus-ar-glasses-smartwatch-plans.

Hecht, Jeff. "Optical Dreams, Virtual Reality." *Optica*, June 2016. https://www.optica -opn.org/home/articles/volume_27/june_2016/features/optical_dreams_virtual _reality/.

Heft-Luthy, Sam. "The Myth of the 'Empathy Machine.'" *Outline*, August 28, 2019. https://theoutline.com/post/7885/virtual-reality-empathy-machine.

Hesch, Joe, Anna Kozminski, and Oskar Linde. "Powered by AI: Oculus Insight." *Meta AI* (blog), August 22, 2019. https://ai.facebook.com/blog/powered-by-ai-oculus -insight/.

Hodge, Stephen L. System and method for personalized virtual reality experience in a controlled environment. US Patent 20,230,015,909, filed August 1, 2022, and issued January 19, 2023. https://patents.justia.com/patent/20230015909.

Hoffmann, Anna Lauren. "Terms of Inclusion: Data, Discourse, Violence." *New Media & Society* 23, no. 12 (2021): 3539–3556. https://doi.org/10.1177/1461444820958725.

Hoffmann, Anna Lauren, Nicholas Proferes, and Michael Zimmer. "'Making the World More Open and Connected': Mark Zuckerberg and the Discursive Construction of Facebook and Its Users." *New Media & Society* 20, no. 1 (2018): 199–218. https://doi.org/10.1177/1461444816660784.

Holler, Drew. "Customers Are Changing. Jobs are Changing. At Walmart, the Future of Work Is Bright." *LinkedIn Pulse* (blog), October 30, 2019. https://www.linkedin .com/pulse/customers-changing-jobs-walmart-future-work-bright-drew-holler?articleId =6595296811185958912.

Hollister, Sean. "BlueShark: Where the US Navy Dreams Up the Battleship Interfaces of Tomorrow." Verge, January 26, 2014. https://www.theverge.com/2014/1/26 /5346772/blueshark-us-navy-oculus-rift-virtual-interface.

Hwang, Tim. *Subprime Attention Crisis: Advertising and the Time Bomb at the Heart of the Internet*. New York: Macmillan, 2020.

IITSEC. "Army Training Mission at I/ITSEC 2021." Recorded panel discussion from the 2021 Interservice/Industry Training, Simulation and Education Conference (IITSEC). YouTube, December 1, 2021. https://www.youtube.com/watch?v=niwpW2qTiEI&ab _channel=IITSEC.

Immersive Wisdom. "Immersive Wisdom Awarded $950MM Ceiling IDIQ Contract to Deliver Its Real-Time JADC2 3D Geospatial Collaboration Software for the US Air Force Advanced Battle Management System (ABMS)." PR Newswire, June 3, 2020. https://www.prnewswire.com/news-releases/immersive-wisdom-awarded-950mm -ceiling-idiq-contract-to-deliver-its-real-time-jadc2-3d-geospatial-collaboration-software -for-the-us-air-force-advanced-battle-management-system-abms-301070226.html.

Immersive Wisdom. "Remote Ops Center." Accessed March 23, 2023. https://www .immersivewisdom.com/.

In-Q-Tel. "Immersive Wisdom Secures Strategic Investment from In-Q-Tel to Scale Its Virtual and Augmented Reality Collaboration and Intelligence Platform." IQT, December 11, 2018. https://www.iqt.org/news/immersive-wisdom-secures-strategic -investment-from-in-q-tel-to-scale-its-virtual-and-augmented-reality-collaboration -and-intelligence-platform/.

Irom, Bimbisar. "Virtual Reality and the Syrian Refugee Camps: Humanitarian Communication and the Politics of Empathy." *International Journal of Communication* 12 (2018): 4269–4291. https://ijoc.org/index.php/ijoc/article/view/8783.

Jackson, Henry, and Jonathan Schenker. "Dealing with Harassment in VR." UploadVR, October 25, 2016. https://uploadvr.com/dealing-with-harassment-in-vr/.

Jensen, Lasse, and Flemming Konradsen. "A Review of the Use of Virtual Reality Head-Mounted Displays in Education and Training." *Education and Information Technologies* 23 (2018): 1515–1529. https://doi.org/10.1007/s10639-017-9676-0.

Jerrett, Adam, Peter Howell, and Neil Dansey. "Developing an Empathy Spectrum for Games." *Games and Culture* 16, no. 6 (2021): 635–659. https://doi.org/10.1177 /1555412020954019.

Jigsaw. "Immersive Technology to Help Advance Public Safety." Jigsaw, Medium, October 26, 2021. https://medium.com/jigsaw/adaptive-technology-to-help-advance -public-safety-b4256388dd3.

Kaplan, Alexandra D., Jessica Cruit, Mica Endsley, Suzanne M. Beers, Ben D. Sawyer, and P. A. Hancock. "The Effects of Virtual Reality, Augmented Reality, and Mixed Reality as Training Enhancement Methods: A Meta-Analysis." *Human Factors: The Journal of the Human Factors and Ergonomics Society* 63, no. 4 (2021): 706–726. https:// doi.org/10.1177/0018720820904229.

Karpf, David. "Virtual Reality Is the Rich White Kid of Technology." *WIRED*, July 27, 2021. https://www.wired.com/story/virtual-reality-rich-white-kid-of-technology/.

Kennedy, Jenny, Michael Arnold, Martin Gibbs, Bjorn Nansen, and Rowan Wilken. *Digital Domesticity: Media, Materiality, and Home Life.* Oxford: Oxford University Press, 2020.

Kilteni, Konstantina, Raphaela Groten, Mel Slater. "The Sense of Embodiment in Virtual Reality." *Presence* 21, no. 4 (2012): 373–387.

Kim, Paulin T. "Data-Driven Discrimination at Work." *William & Mary Law Review* 58, no. 3 (2017): 857–936. https://scholarship.law.wm.edu/wmlr/vol58/iss3/4.

Kitchin, Rob. *The Data Revolution: Big Data, Open Data, Data Infrastructures and Their Consequences.* New York: SAGE, 2014.

Kocurek, Carly A. *Coin-Operated Americans: Rebooting Boyhood at the Video Game Arcade.* Minneapolis: University of Minnesota Press, 2015.

Koenecke, Allison, Andrew Nam, Emily Lake, Joe Nudell, Minnie Quartey, Zion Mengesha, Connor Toups, John R. Rickford, Dan Jurafsky, and Sharad Goel. "Racial Disparities in Automated Speech Recognition." In *Proceedings of the National Academy of Sciences* 117, no. 14 (2020): 7684–7689. https://doi.org/10.1073/pnas.1915768117.

Kratos Defense & Security Solutions. "Simulation Training Platform." Accessed March 23, 2023. https://www.kratosdefense.com/systems-and-platforms/training-systems/simulation-training-platform.

Kuchera, Ben. "Superhot VR Is a Whole New Game, Built for the Oculus Touch." Polygon, October 10, 2016. https://www.polygon.com/2016/10/10/13227564/superhot-vr-oculus-rift-impressions.

Kumparak, Greg. "A Brief History of Oculus." TechCrunch, March 27, 2014. https://techcrunch.com/2014/03/26/a-brief-history-of-oculus/.

Kutska, Debra. "Variation in Visitor Perceptions of a Polar Bear Enclosure Based on the Presence of Natural vs. Un-natural Enrichment Items." *Zoo Biology* 28, no. 4 (2009): 292–306.

Lai, Jonathan. "Meet Me in the Metaverse." Andreessen Horowitz, December 7, 2020. https://a16z.com/2020/12/07/social-strikes-back-metaverse/.

Lam, Joseph. "VR Group Lands $1.7m WA Police Deal." *Australian*, November 7, 2022. https://www.theaustralian.com.au/web-stories/free/the-australian/vr-group-lands-1-7m-wa-police-deal.

Lanier, Jaron. *Dawn of the New Everything: A Journey through Virtual Reality*. London: Bodley Head, 2017.

Leibovitz, Liel. *God in the Machine: Video Games as Spiritual Pursuit*. West Conshohocken, PA: Templeton Press, 2014.

Levy, Steven. "The Metaverse Is Mark Zuckerberg's Mobile Do-Over." *WIRED*, November 5, 2021. https://www.wired.com/story/plaintext-metaverse-mark-zuckerberg-mobile-do-over/.

Lien, Tracey. "No Girls Allowed." Polygon, December 2, 2013. https://www.polygon.com/features/2013/12/2/5143856/no-girls-allowed.

Luckey, Palmer. "Update #36: VR Gets VC." Oculus Rift: Step Into the Game, crowdfunding project, Kickstarter, June 18, 2013. https://www.kickstarter.com/projects/1523379957/oculus-rift-step-into-the-game/posts/512945.

Maloney, Divine, Guo Freeman, and Andrew Robb. "It Is Complicated: Interacting with Children in Social Virtual Reality." in *2020 IEEE Conference on Virtual Reality and 3D User Interfaces Abstracts and Workshops*, 343–347. Piscataway, NJ: IEEE, 2020. https://doi.org/10.1109/VRW50115.2020.00075.

Maloney, Divine, Guo Freeman, and Andrew Robb. "A Virtual Space for All: Exploring Children's Experience in Social Virtual Reality." In *CHI Play '20: Proceedings of the*

Annual Symposium on Computer-Human Interaction in Play, 472–483. New York: Association for Computing Machinery, 2020. https://doi.org/10.1145/3410404.3414268.

Marcus, George E. *Technoscientific Imaginaries*. Chicago: Chicago University Press, 1994.

Matney, Lucas. "Oculus Research Is Now the 'Facebook Reality Lab.'" TechCrunch, May 8, 2018. https://techcrunch.com/2018/05/07/oculus-research-is-now-the-facebook-reality-lab/.

Matsuda, Keichii. "Hyper-Reality." YouTube, May 20, 2016. https://www.youtube.com/watch?v=YJg02ivYzSs.

McFerran, Damien. "Reality Crumbles: Whatever Happened to VR?" Eurogamer, March 23, 2014. https://www.eurogamer.net/reality-crumbles-whatever-happened-to-vr.

McQuillan, Dan. *Resisting AI: An Anti-Fascist Approach to Artificial Intelligence*. Bristol: Bristol University Press, 2022.

Mehrabi, Ninareh, Fred Morstatter, Nripsuta Saxena, Kristina Lerman, and Aram Galstyan. "A Survey on Bias and Fairness in Machine Learning." *ACM Computing Surveys* 54, no. 6, article 115 (2021): 1–35. https://doi.org/10.1145/3457607.

Meta. "First Quarter 2022 Results Conference Call." Meta, April 27, 2022. https://s21.q4cdn.com/399680738/files/doc_financials/2022/q1/Meta-Q1-2022-Earnings-Call-Transcript.pdf.

Meta Quest. "Day 1 Keynote | Oculus Connect 6." Recorded talk from the 2019 Oculus Connect 6 Developer Conference. YouTube, September 25, 2019. https://www.youtube.com/watch?v=RCB_mfGmh9w&ab_channel=MetaQuest.

Meta Quest. "Horizon Worlds." YouTube, October 7, 2021. https://www.youtube.com/watch?v=PwVV-zczHYI&ab_channel=MetaQuest.

Meta Quest. "Michael Abrash Keynote from F8 2015." Recorded talk from the 2015 Facebook Developer Conference (F8), March 25–26, 2015, Fort Mason Center, San Francisco. YouTube, May 8, 2015. https://www.youtube.com/watch?v=XVCthGEFwHw&ab_channel=MetaQuest.

Meta Quest. "Oculus Connect 4 | Day 1 Keynote." Recorded talk from the 2017 Oculus Connect 4 Developer Conference. YouTube, October 13, 2017. https://www.youtube.com/watch?v=QAa1GjiLktc&ab_channel=MetaQuest.

Meta Quest. "Oculus Connect 5 | Keynote Day 02." Recorded talk from the 2018 Oculus Connect 5 Developer Conference. YouTube, September 28, 2018. https://www.youtube.com/watch?v=VW6tgBcN_fA&ab_channel=MetaQuest.

Meta Quest. "Oculus Connect Keynote: Michael Abrash." Recorded talk from the 2014 Oculus Connect 1 Developer Conference. YouTube, October 2, 2014. https://www.youtube.com/watch?v=knQSRTApNcs&ab_channel=MetaQuest.

Metaverse: Another Cesspool of Toxic Content." SumOfUs, May 2022. https://www.eko
.org/images/Metaverse_report_May_2022.pdf.

Milk, Chris. "How Virtual Reality Can Create the Ultimate Empathy Machine."
Filmed at TED2015. TED Video, March 2015. https://www.ted.com/talks/chris_milk
_how_virtual_reality_can_create_the_ultimate_empathy_machine?language=en.

Miller, Mark Roman, Fernanda Herrera, Hanseul Jun, James A. Landay, and Jeremy
N. Bailenson. "Personal Identifiability of User Tracking Data during Observation
of 360-Degree VR Video." *Scientific Reports* 10, no. 1 (2020): 1–10. https://doi.org
/10.1038/s41598-020-74486-y.

Miller, Robert, Natasha Kholgade Banerjee, and Sean Banerjee. "Within-System and
Cross-System Behavior-Based Biometric Authentication in Virtual Reality." In *Proceedings of the 2020 IEEE Conference on Virtual Reality and 3D User Interfaces Abstracts
and Workshops*, 311–316. Piscataway, NJ: IEEE, 2020. https://doi.org/10.1109/VRW
50115.2020.00070.

Mir, Roy, and Katitza Rodriguez. "If Privacy Dies in VR, It Dies in Real Life." *Deep-
links Blog*, Electronic Frontier Foundation, August 25, 2020. https://www.eff.org
/deeplinks/2020/08/if-privacy-dies-vr-it-dies-real-life.

Morrissette, Jess. "How Games Marketing Invented Toxic Gamer Culture." *VICE*,
March 25, 2020. https://www.vice.com/en/article/5dmayn/games-marketing-toxic
-gamer-culture-online-xbox-live-dreamcast.

Munafo, Justin, Meg Diedrick, and Thomas A. Stoffregen. "The Virtual Reality
Head-Mounted Display Oculus Rift Induces Motion Sickness and Is Sexist in Its
Effects." *Experimental Brain Research* 235 (2017): 889–901. https://doi.org/10.1007
/s00221-016-4846-7.

Murphy, Hannah. "How Will Facebook Keep Its Metaverse Safe for Users?" *Financial
Times*, November 12, 2021. https://www.ft.com/content/d72145b7-5e44-446a-819c
-51d67c5471cf.

Murray, Janet H. "Not a Film and Not an Empathy Machine." Immerse, Medium,
October 7, 2016. https://immerse.news/not-a-film-and-not-an-empathy-machine
-48b63b0eda93.

Nakamura, Lisa. "Feeling Good about Feeling Bad: Virtuous Virtual Reality and the
Automation of Racial Empathy." *Journal of Visual Culture* 19, no 1. (2020): 47–64.

NASA Spinoff. "Computer Technology: A New Continent of Ideas." *Spinoff*, January
1, 1990, 88–91. http://hdl.handle.net/hdl:2060/20020086961.

Nash, Kate. "Virtual Reality Witness: Exploring the Ethics of Mediated Presence."
Studies in Documentary Film 12, no. 2 (2018): 119–131. https://doi.org/10.1080/1750
3280.2017.1340796.

Newn, Joshua, Fraser Allison, Eduardo Velloso, and Frank Vetere. "Looks Can Be
Deceiving: Using Gaze Visualisation to Predict and Mislead Opponents in Strategic

Gameplay." In *Proceedings of the 2018 CHI Conference on Human Factors in Computing Systems*, 1–12. New York: Association for Computing Machinery, 2018. https://doi .org/10.1145/3173574.3173835.

Nieborg, David B., and Anne Helmond. "The Political Economy of Facebook's Platformization in the Mobile Ecosystem: Facebook Messenger as a Platform Instance." *Media, Culture & Society* 41, no. 2 (2019): 196–218. https://doi.org/10.1177 /0163443718818384.

Nix, Naomi, and Mark Gurman. "Meta's Oculus Unit Faces FTC-Led Probe of Competition Practices." Bloomberg, January 15, 2022. https://www.bloomberg.com /news/articles/2022-01-14/meta-s-oculus-unit-faces-ftc-led-probe-of-competition -practices#xj4y7vzkg.

Noble, Safiya Umoja. *Algorithms of Oppression*. New York: New York University Press, 2018.

Nordmann, Alfred. "If and Then: A Critique of Speculative NanoEthics." *NanoEthics* 1, no. 31 (2007): 31–46. https://doi.org/10.1007/s11569-007-0007-6.

Novet, Jordan. "Microsoft Wins U.S. Army Contract for Augmented Reality Headsets, Worth up to $21.9 Billion over 10 Years." CNBC, last updated April 1, 2021. https://www.cnbc.com/2021/03/31/microsoft-wins-contract-to-make-modified -hololens-for-us-army.html.

Oculus VR. "Introducing App Lab: A New Way to Distribute Oculus Quest Apps." *Meta Quest Developer Center* (blog), last updated November 16, 2021. https://developer .oculus.com/blog/introducing-app-lab-a-new-way-to-distribute-oculus-quest-apps/.

Oculus VR. "VR for Good: Creators Lab Returns for 2017." *Meta Quest Blog*, Meta, May 8, 2017. https://www.meta.com/blog/quest/vr-for-good-creators-lab-returns-for-2017/.

O'Dell, J. "Facebook Is Buying Oculus for $2B to 'Get Ready for the Platform of Tomorrow.'" VentureBeat, March 25, 2014. https://venturebeat.com/entrepreneur /facebook-is-buying-oculus-for-2b-to-get-ready-for-the-platform-of-tomorrow/.

Open Innovation Team. "Investigating Virtual Reality for the College of Policing." Gov.UK, July 22, 2022. https://www.gov.uk/government/news/investigating-virtual -reality-for-the-college-of-policing.

Oremus, Will. "In 2021, Tech Talked up 'the Metaverse.' One Problem: It Doesn't Exist." *Washington Post*, December 30, 2021. https://www.washingtonpost.com /technology/2021/12/30/metaverse-definition-facebook-horizon-worlds/.

Orland, Kyle. "Rift Goes Wireless: Ars Walks around in Oculus' Santa Cruz VR Prototype." *ArsTechnica*, October 8, 2016. https://arstechnica.com/gaming/2016/10/rift -goes-wireless-ars-walks-around-in-oculus-santa-cruz-vr-prototype/.

Outlaw, Jessica. "Virtual Harassment: The Social Experience of 600+ Regular Virtual Reality (VR) Users." Virtual Reality Pop, Medium, April 4, 2018. https://virtualreali

typop.com/virtual-harassment-the-social-experience-of-600-regular-virtual-reality
-vr-users-23b1b4ef884e.

Pahwa, Nitish. "Why the Metaverse Has to Look So Stupid." *Slate*, August 19, 2022.
https://slate.com/technology/2022/08/mark-zuckerberg-metaverse-horizon-worlds
-facebook-looks-crappy-explained.html.

Parisi, David. *Archaeologies of Touch: Interfacing with Haptics from Electricity to Computing*. Minneapolis: University of Minnesota Press, 2018.

Parshall, Allison. "Why an Assault on Your VR Body Can Feel So Real." Scienceline,
June 29, 2022. https://scienceline.org/2022/06/virtual-reality-assault-psychology/.

Partin, William Clyde. "Bit by (Twitch) Bit: 'Platform Capture' and the Evolution
of Digital Platforms." *Social Media + Society* 6, no. 3 (2020): 1–12. https://doi.org
/10.1177/2056305120933981.

Paterson, Mark. "On Haptic Media and the Possibilities of a More Inclusive Interactivity." *New Media & Society* 19, no. 10 (2017): 1541–1562. https://doi.org/10.1177
/1461444817717513.

Peck, Tabitha C., Laura E. Sockol, and Sarah M. Hancock. "Mind the Gap: The Under-
representation of Female Participants and Authors in Virtual Reality Research." *IEEE
Transactions on Visualization and Computer Graphics* 26, no. 5 (2020): 1945–1954.
https://doi.org/10.1109/TVCG.2020.2973498.

Pesce, Mark. *Augmented Reality: Unboxing Tech's Next Big Thing*. Cambridge: Polity,
2021.

Pfeuffer, Ken, Matthias J. Geiger, Sarah Prange, Lukas Mecke, Daniel Buschek, and
Florian Alt. "Behavioural Biometrics in VR: Identifying People from Body Motion
and Relations in Virtual Reality." In *Proceedings of the 2019 CHI Conference on Human
Factors in Computing Systems*, 1–12. New York: Association for Computing Machinery, 2019. https://doi.org/10.1145/3290605.3300340.

Phan, Thao, Jake Goldenfein, Monique Mann, and Declan Kuch. "Economies of
Virtue: The Circulation of 'Ethics' in Big Tech." *Science as Culture* 31, no. 1 (2022):
121–135. https://doi.org/10.1080/09505431.2021.1990875.

Pike, John. "'Something That Is Ours': VR and the Values of Gaming's Field." In
Proceedings of DiGRA Australia, 1–2. Pittsburgh: ETC Press, 2019. https://digraa.org
/wp-content/uploads/2019/01/DIGRAA_2019_paper_31.pdf.

Raghavan, Manish, Solon Barocas, Jon Kleinberg, and Karen Levy. "Mitigating Bias
in Algorithmic Hiring: Evaluating Claims and Practices." In *Proceedings of the 2020
Conference on Fairness, Accountability, and Transparency*, 469–481. New York: Association for Computing Machinery, 2020. https://doi.org/10.1145/3351095.3372828.

Rawski, Shannon L., Joshua R. Foster, and Jeremy Bailenson. "Sexual Harassment
Bystander Training Effectiveness: Experimentally Comparing 2D Video to Virtual

Reality Practice." *Technology, Mind, and Behavior* 3, no. 2 (2022): 1–51. https://doi.org /10.1037/tmb0000074.

Raytheon Intelligence and Space. "Synthetic Training Environment." Accessed March 23, 2023. https://www.raytheonintelligenceandspace.com/what-we-do/advanced-tech /synthetic-environments/synthetic-training-environment.

Reality Labs. "From the Lab to the Living Room: The Story Behind Facebook's Oculus Insight Technology and a New Era of Consumer VR." Meta, August 21, 2019. https:// tech.facebook.com/reality-labs/2019/8/the-story-behind-oculus-insight-technology/.

Reality Labs. "Inside Facebook Reality Labs: Wrist-Based Interaction for the Next Computing Platform." Meta, March 18, 2021. https://tech.facebook.com/reality -labs/2021/03/inside-facebook-reality-labs-wrist-based-interaction-for-the-next -computing-platform/.

Reality Labs. "Project Aria: Facebook Connect." Facebook, September 16, 2020. https://www.facebook.com/watch/?v=2616922421880787.

Reuters, "You're Fired! Meet Barry—the VR Character You Can Fire Again and Again." Reuters—Screen Ocean, September 2019.

Rheingold, Howard. *Virtual Reality: The Revolutionary Technology of Computer-Generated Artificial Worlds—and How It Promises to Transform Society*. New York: Touchstone, 1991.

Rivero, Nicolás. "Consulting Firms Are Betting on Unproven VR Diversity Trainings." Quartz, June 25, 2020. https://qz.com/1872927/do-vr-diversity-trainings-work.

Robb, Lori A., and Dennis Doverspike. "Self-Reported Proclivity to Harass as a Moderator of the Effectiveness of Sexual Harassment-Prevention Training." *Psychological Reports* 88, no. 1 (2001), 85–88. https://doi.org/10.2466/pr0.2001.88.1.85.

Robertson, Adi. "Building for Virtual Reality? Don't Forget about Women." Verge, January 11, 2016. https://www.theverge.com/2016/1/11/10749932/vr-hardware -needs-to-fit-women-too.

Robertson, Adi. "Facebook Is turning VR into a Platform—but Some Indie Developers Fear Its Power." Verge, September 25, 2020. https://www.theverge.com/21455665 /facebook-oculus-vr-indie-developers-power-monopoly-concerns.

Robertson, Adi, and Michael Zelenko. "Voices from a Virtual Past." Verge. Accessed March 23, 2023. https://www.theverge.com/a/virtual-reality/oral_history.

Robinson, Joanna. "The Sci-Fi Guru Who Predicted Google Earth Explains Silicon Valley's Latest Obsession." *Vanity Fair*, June 23, 2017. https://www.vanityfair.com /news/2017/06/neal-stephenson-metaverse-snow-crash-silicon-valley-virtual-reality.

Rodriguez, Salvador. "Facebook's Meta Mission Was Laid Out in a 2018 Paper Declaring 'The Metaverse Is Ours to Lose.'" CNBC, October 30, 2021. https://www

.cnbc.com/2021/10/30/facebooks-meta-mission-was-laid-out-in-a-2018-paper-on-the
-metaverse.html.

"Roger Ebert on Empathy." Video featuring Roger Ebert's speech on empathy, delivered in July 2005 at the dedication of his plaque outside the Chicago Theatre. RogerEbert.com, April 4, 2018. https://www.rogerebert.com/empathy/video-roger -ebert-on-empathy.

Roose, Kevin. "The Metaverse Is Mark Zuckerberg's Escape Hatch." *New York Times*, October 29, 2021. https://www.nytimes.com/2021/10/29/technology/meta-facebook -zuckerberg.html.

Rose, Mandy. "The Immersive Turn: Hype and Hope in the Emergence of Virtual Reality as a Nonfiction Platform." *Studies in Documentary Film* 12, no. 2 (2018): 132– 149. https://doi.org/10.1080/17503280.2018.1496055.

Rosenblat, Alex. *Uberland: How Algorithms Are Rewriting the Rules of Work.* Oakland: University of California Press, 2019.

Rubin, Peter. "The Inside Story of Oculus Rift and How Virtual Reality Became Reality." *WIRED*, May 20, 2014. https://www.wired.com/2014/05/oculus-rift-4/.

Saab (@Saab). "The secondary trigger in last tweet is seen on the Carl-Gustaf M4 VR Experience, an opportunity for visitors at #DSEI to try. Welcome!" Twitter, September 13, 2017, 5:18 a.m. https://twitter.com/Saab/status/907941588436963329.

Sadowski, Jathan. "Future Schlock." *Real Life Magazine*, January 25, 2021. https:// reallifemag.com/future-schlock/.

Sadowski, Jathan. *Too Smart: How Digital Capitalism Is Extracting Data, Controlling Our Lives, and Taking Over the World.* Cambridge, MA: MIT Press, 2020.

Sadowski, Jathan. "When Data Is Capital: Datafication, Accumulation, and Extraction." *Big Data & Society* 6, no. 1 (2019): 1–12. https://doi.org/10.1177/205395171 8820549.

Sadowski, Jathan, and Kaitlin Beegle. "Expansive and Extractive Networks of Web3." *Big Data & Society* 10, no. 1 (2023): 1–14. https://doi.org/10.1177/20539517231159629.

Saker, Michael, and Jordan Frith. "Coextensive Space: Virtual Reality and the Developing Relationship between the Body, the Digital and Physical Space." *Media, Culture & Society* 42, no. 7–8 (2020): 1427–1442. https://doi.org/10.1177/0163443720932498.

Sánchez-Monedero, Javier, Lina Dencik, and Lilian Edwards. "What Does It Mean to 'Solve' the Problem of Discrimination in Hiring? Social, Technical and Legal Perspectives from the UK on Automated Hiring Systems." In *Proceedings of the 2020 Conference on Fairness, Accountability, and Transparency*, 458–468. New York: Association for Computing Machinery, 2020. https://doi.org/10.1145/3351095.3372849.

Savage, Adam. "Inside Valve: Making *Half-Life: Alyx* for Virtual Reality." An in-depth interview with Robin Walker and Greg Coomer about designing *Half-Life: Alyx*.

Adam Savage's Tested, YouTube, March 5, 2020. https://www.youtube.com/watch ?v=cRVXhA0-TI4.

Scharre, Paul. "How Swarming Will Change Warfare." *Bulletin of the Atomic Scientists* 74, no. 6 (2018): 385–389. https://doi.org/10.1080/00963402.2018.1533209.

Schneider, Nathan. "Decentralization: An Incomplete Ambition." *Journal of Cultural Economy* 12, no. 4 (2019): 265–285. https://doi.org/10.1080/17530350.2019.1589553.

Selwyn, Neil. "Do(n't) Believe the Hype . . . Engaging with Tech Industry Expectations of Emerging Technologies." *Critical Studies of Education and Technology* (blog), June 25, 2021. https://criticaledtech.com/2021/06/25/dont-believe-the-hype-engag ing-with-tech-industry-expectations-of-emerging-technologies/.

Seufert, Eric (@eric_seufert). "I dislike using 'metaverse' to describe what Facebook is doing because it's unhelpful jargon, but what I think FB wants to build is: a persistent economy of . . ." Twitter, July 29, 2021, 8:24 a.m. https://twitter.com/eric _seufert/status/1420767147278716934.

SIGGRAPH Conferences. "VR @ 50: Ivan Sutherland's 1968 Head-Mounted 3D Display System." *ACM SIGGRAPH Blog*, August 14, 2018. https://blog.siggraph.org /2018/08/vr-at-50-celebrating-ivan-sutherland.html/.

Silberling, Amanda. "FTC Puts Zuckerberg on the Stand over Meta's Plan to Acquire Within." TechCrunch, December 20, 2022. https://techcrunch.com/2022/12/20/ftc -puts-zuckerberg-on-the-stand-over-metas-plan-to-acquire-within/.

Silverman, Dwight. "NASA Explores Virtual Reality." *Houston Chronicle*, June 7, 1992. https://www.houstonchronicle.com/techburger/article/From-the-archives-Virtual -Reality-12709811.php.

Simon, Bart. *Wii Are Out of Control: Bodies, Game Screens and the Production of Gestural Excess.* Social Science Research Network (SSRN), March 5, 2009. http://dx.doi .org/10.2139/ssrn.1354043.

Simonite, Tom. "Behind Anduril's Effort to Create an Operating System for War." *WIRED*, October 8, 2020. https://www.wired.com/story/behind-anduril-effort-create -operating-system-war/.

Slater, Mel, Vasilis Linakis, Martin Usoh, and Rob Kooper. "Immersion, Presence and Performance in Virtual Environments: An Experiment with Tri-Dimensional Chess." In *Proceedings of the ACM Symposium on Virtual Reality Software and Technology*, 163–172. New York: Association for Computing Machinery, 1996. https://doi .org/10.1145/3304181.3304216.

Solon, Olivia. "Mark Zuckerberg 'Tours' Flooded Puerto Rico in Bizarre Virtual Reality Promo." *Guardian*, October 9, 2017. https://www.theguardian.com/technology /2017/oct/09/mark-zuckerberg-facebook-puerto-rico-virtual-reality.

Sophia, Zoë. "Virtual Corporeality: A Feminist View." *Australian Feminist Studies* 7, no. 15 (1992): 11–24. https://doi.org/10.1080/08164649.1992.9994641.

Stanney, Kay, Cali Fidopiastis, and Linda Foster. "Virtual Reality Is Sexist: But It Does Not Have to Be." *Frontiers in Robotics and AI* 7, article 4 (2020): 1–19. https://doi.org/10.3389/frobt.2020.00004.

Statt, Nick. "Facebook Reportedly Ignored Its Own Research Showing Algorithms Divided Users." Verge, May 26, 2020. https://www.theverge.com/2020/5/26/2127 0659/facebook-division-news-feed-algorithms.

Stilgoe, Jack, Richard Owen, and Phil Macnaghten. "Developing a Framework for Responsible Innovation." *Research Policy* 42, no. 9 (2013): 1568–1580. https://doi.org /10.1016/j.respol.2013.05.008.

Suits, Devon L. "Synthetic Training Environment to Enhance Soldier Lethality." US Army, March 28, 2018. https://www.army.mil/article/202574/synthetic_training _environment_to_enhance_soldier_lethality.

Suso-Ribera, Carlos, Javier Fernández-Álvarez, Azucena García-Palacios, Hunter G. Hoffman, Juani Bretón-López, Rosa M. Banos, Soledad Quero, and Cristina Botella. "Virtual Reality, Augmented Reality, and *In Vivo* Exposure Therapy: A Preliminary Comparison of Treatment Efficacy in Small Animal Phobia." *Cyberpsychology, Behavior, and Social Networking* 22, no. 1 (2019): 31–38. https://doi.org/10.1089/cyber .2017.0672.

Sutherland, Ivan E. "A Head-Mounted Three-Dimensional Display." In *Proceedings of the December 9–11, 1968, Fall Joint Computer Conference, Part 1*, 757–764. New York: Association for Computing Machinery, 1968. https://doi.org/10.1145/1476589 .1476686.

Swauger, Shea. "Our Bodies Encoded: Algorithmic Test Proctoring in Higher Education." *Hybrid Pedagogy*, April 2, 2020. https://hybridpedagogy.org/our-bodies -encoded-algorithmic-test-proctoring-in-higher-education/.

Táíwò, Olúfẹ́mi O. *Elite Capture: How the Powerful Took Over Identity Politics (and Everything Else)*. Chicago: Haymarket Books, 2022.

Takahashi, Dean. "How VR Can Help Enterprises with Training, beyond Firing Barry." Venturebeat, September 29, 2019. https://venturebeat.com/2019/09/29/how -vr-can-help-enterprises-with-training-beyond-firing-barry/.

Tarnoff, Ben. *Internet for the People: The Fight for Our Digital Future*. New York: Verso Books, 2022.

Thomas, Sarah. *Social Change for Conservation: The World Zoo and Aquarium Conservation Education Strategy*. Barcelona: World Association of Zoos and Aquariums Executive Office, October 2020. https://www.waza.org/wp-content/uploads/2020/10 /10.06_WZACES_spreads_20mbFINAL.pdf.

Thomson, Rosemary Garland. *Extraordinary Bodies: Figuring Physical Disability in American Culture and Literature*. New York: Columbia University Press, 1997.

Trimananda, Rahmadi, Hieu Le, Hao Cui, and Janice Tran Ho. "OVRseen: Auditing Network Traffic and Privacy Policies in Oculus VR." In *Proceedings of the 31st USENIX Security Symposium*, 3789–3806. Berkeley, CA: USENIX Association, 2022.

US Equal Employment Opportunity Commission. "Task Force Co-Chairs Call on Employers and Others to 'Reboot' Harassment Prevention." Press release, June 20, 2016. https://www.eeoc.gov/newsroom/task-force-co-chairs-call-employers-and-others -reboot-harassment-prevention?renderforprint=1.

United Nations SDG Action Campaign. "Syrian Refugee Crisis: Clouds over Sidra." Accessed March 23, 2023. http://unvr.sdgactioncampaign.org/cloudsoversidra/#.ZB y9cnZByUl.

United States Army Acquisition Support Center. "Synthetic Training Environment (STE)." Accessed March 23, 2023. https://asc.army.mil/web/portfolio-item/synthetic -training-environment-ste/.

UploadVR. "Andrew Bosworth OC6: AR Glasses, Social VR & Live Maps". Recorded talk from the 2019 Oculus Connect 6 Conference. YouTube, September 25, 2019. https://www.youtube.com/watch?v=0woRq7EYZ2Q&ab_channel=UploadVR.

UploadVR. "Michael Abrash OC6: The Future of VR." Recorded talk from the 2019 Oculus Connect 6 Conference. YouTube, September 25, 2019. https://www.youtube .com/watch?v=7YIGT13bdXw&ab_channel=UploadVR.

Urry, John. "The 'Consumption' of Tourism." *Sociology* 24, no.1 (1990): 23–35. https://doi.org/10.1177/0038038590024001004.

Valve. "Best of 2022—Best of VR." Steam. Accessed March 23, 2023. https://store .steampowered.com/sale/BestOf2022?tab=5.

Vaselli, Francesca Liberatore. "Changing Humanitarianism for the Better? Virtual Reality and the Representation of the Suffering 'Other' in Humanitarian Communications." Master's diss., University of London, 2021. https://www.lse.ac.uk/media -and-communications/assets/documents/research/msc-dissertations/2020/248 -Vaselli.pdf.

Vinciguerra, Robert A. "Tom Kalinske Talks about His Time Overseeing Sega as Its CEO in the 90s; Reveals That Sega Passed on Virtual Boy Technology, Considered Releasing 3DO." *Rev. Rob Times* (blog), September 21, 2015. https://web.archive.org /web/20150924024256/http:/revrob.com/sci-tech/264-tom-kalinske-talks-about-his -time-overseeing-sega-as-its-ceo-in-the-90s-reveals-that-sega-passed-on-virtual-boy -technology-considered-releasing-3do.

Virilio, Paul, and Sylvere Lotringer. *Pure War*. Los Angeles: Semiotext(e), 1998.

Wang, Jackie. *Carceral Capitalism*. Cambridge, MA: MIT Press, 2018.

Warzel, Charlie. "Is QAnon the Most Dangerous Conspiracy Theory of the 21st Century?" *New York Times*, August 4, 2020. https://www.nytimes.com/2020/08/04 /opinion/qanon-conspiracy-theory-arg.html.

Webber, Sarah, Marcus Carter, Sally Sherwen, Wally Smith, Zaher Joukhadar, and Frank Vetere. "Kinecting with Orangutans: Zoo Visitors' Empathetic Responses to Animals' Use of Interactive Technology." In *Proceedings of the 2017 CHI Conference on Human Factors in Computing Systems*, 6075–6088. New York: Association for Computing Machinery, 2017. https://doi.org/10.1145/3025453.3025729.

Webber, Sarah, Marcus Carter, Wally Smith, and Frank Vetere. "Interactive Technology and Human–Animal Encounters at the Zoo." *International Journal of Human-Computer Studies* 98 (2017): 150–168. https://doi.org/10.1016/j.ijhcs.2016.05.003.

Welsh, Oli. "John Carmack and the Virtual Reality Dream." Eurogamer, June 7, 2012. https://www.eurogamer.net/john-carmack-and-the-virtual-reality-dream.

White House. "Fact Sheet: President Biden Sends Immigration Bill to Congress as Part of His Commitment to Modernize Our Immigration System." Press release, January 20, 2021. https://www.whitehouse.gov/briefing-room/statements-releases /2021/01/20/fact-sheet-president-biden-sends-immigration-bill-to-congress-as-part -of-his-commitment-to-modernize-our-immigration-system/.

Whitehouse, Rich. "Sega VR Revived: Emulating an Unreleased Genesis Accessory." *Video Game History Foundation* (blog), November 20, 2020. https://gamehistory.org /segavr/.

Winner, Langdon. "Do Artifacts Have Politics?" *Daedelus* 109, no. 1 (1980): 121–136. https://www.jstor.org/stable/20024652.

Won, Andrea Stevenson, Jeremy N. Bailenson, and Joris H. Janssen. "Automatic Detection of Nonverbal Behavior Predicts Learning in Dyadic Interactions." *IEEE Transactions on Affective Computing* 5, no. 2 (2014): 112–125. https://doi.org/10.1109 /TAFFC.2014.2329304.

Woolley, Benjamin. *Virtual Worlds: A Journey in Hype and Hyperreality*. Oxford: Blackwell Publishing, 1992.

Woputz, Cody. "How Giving Mark Zuckerberg a Demo Changed My View of VR." MKAI, June 29, 2022. https://mkai.org/how-giving-mark-zuckerberg-a-demo -changed-my-view-of-vr/?utm_source=rss&utm_medium=rss&utm_campaign=how -giving-mark-zuckerberg-a-demo-changed-my-view-of-vr.

World Economic Forum. "Defining and Building the Metaverse." Accessed March 24, 2023. https://initiatives.weforum.org/defining-and-building-the-metaverse/home.

Zhangshao, Tianyi, and Marcus Carter. "A Hybrid Revolution: The Appeal of Hybrid Gaming on the Nintendo Switch." In *Proceedings of the 57th Annual Hawaii International Conference on System Sciences*, 2654–2664. Honolulu: HICSS, 2024.

Zimmerman, Thomas G., Jaron Lanier, Chuck Blanchard, Steve Bryson, and Young Harvill. "A Hand Gesture Interface Device." *ACM SIGCHI Bulletin* 17, SI (1987): 189–192. https://doi.org/10.1145/30851.275628.

Zitron, Ed. "Mark Zuckerberg Is a Liar, and He's Lying to You about the Metaverse." *Ed Zitron's Where's Your Ed At* (blog), February 8, 2022. https://ez.substack.com/p /mark-zuckerberg-is-a-liar-and-hes.

Zuboff, Shoshana. *The Age of Surveillance Capitalism: The Fight for a Human Future at the New Frontier of Power.* London: Profile Books, 2019.

Zuckerberg, Mark. "2015 06 22 Mark's Vision." Email sent June 22, 2015; document uploaded to Scribd from TechCrunch. Accessed March 23, 2023. https://www.scribd .com/document/399594551/2015-06-22-MARK-S-VISION#from_embed.

Zuckerberg, Mark. "Founders Letter, 2021." Meta, October 28, 2021. https://about.fb .com/news/2021/10/founders-letter/.

Zuckerberg, Mark. "I'm excited to announce that we've agreed to acquire Oculus VR, the leader in virtual reality technology." Facebook, March 25, 2014. https://www .facebook.com/zuck/posts/10101319050523971?stream_ref=10.

INDEX

Page numbers in italics refer to figures.